Lecture Notes in Computer Science 2334

Edited by G. Goos, J. Hartmanis, and J. van Leeuwen

Springer

Berlin
Heidelberg
New York
Barcelona
Hong Kong
London
Milan
Paris
Tokyo

Georg Carle   Martina Zitterbart (Eds.)

Protocols for High Speed Networks

7th IFIP/IEEE International Workshop, PfHSN 2002
Berlin, Germany, April 22-24, 2002
Proceedings

Springer

Series Editors

Gerhard Goos, Karlsruhe University, Germany
Juris Hartmanis, Cornell University, NY, USA
Jan van Leeuwen, Utrecht University, The Netherlands

Volume Editors

Georg Carle
Fraunhofer Institut FOKUS
Kaiserin-Augusta-Allee 31, 10589 Berlin, Germany
E-mail: carle@fokus.gmd.de

Martina Zitterbart
University of Karlsruhe
Faculty of Computer Science, Institute of Telematics
Zirkel 2, 76128 Karlsruhe, Germany
E-mail: zit@tm.uka.de

Cataloging-in-Publication Data applied for

Die Deutsche Bibliothek - CIP-Einheitsaufnahme

Protocols for high speed networks : 7th IFIP/IEEE international workshop ; proceedings / PfHSN 2002, Berlin, Germany, April 22 - 24, 2002. Georg Carle ; Martina Zitterbart (ed.). - Berlin ; Heidelberg ; New York ; Barcelona ; Hong Kong ; London ; Milan ; Paris ; Tokyo : Springer, 2002   (Lecture notes in computer science ; Vol. 2334)   ISBN 3-540-43658-8

CR Subject Classification (1998): C.2, D.4.4, H.3.5, K.4.4

ISSN 0302-9743
ISBN 3-540-43658-8 Springer-Verlag Berlin Heidelberg New York

Springer-Verlag Berlin Heidelberg New York
a member of BertelsmannSpringer Science+Business Media GmbH

http://www.springer.de

©2002 IFIP International Federation for Information Processing, Hofstrasse 3, A-2361 Laxenburg,Austria

Typesetting: Camera-ready by author, data conversion by Steingräber Satztechnik GmbH, Heidelberg
Printed on acid-free paper   SPIN 10869765   06/3142   5 4 3 2 1 0

Preface

This workshop on "Protocols for High-Speed Networks" is the seventh in a successful series of international workshops, well known for their small and focused target audience, that provide a sound basis for intensive discussions of high-quality and timely research work.

The location of the workshop has alternated between Europe and the United States, at venues not only worth visiting for the workshop, but also for the distinct impressions they leave on the participants. The first workshop was held in 1989 in Zurich. Subsequently the workshop was moved to Palo Alto (1990), Stockholm (1993), Vancouver (1994), Sophia-Antipolis/Nice (1996), and Salem (1999). In 2002, the workshop was hosted in Berlin, the capital of Germany.

PfHSN is a workshop providing an international forum that focuses on issues related to high-speed networking, such as protocols, implementation techniques, router design, network processors and the like. Although the topics have shifted during the last couple of years, for example, from parallel protocol implementations to network processors, it could be observed that high speed remains a very important issue with respect to future networking. Traditionally, PfHSN is a relatively focused and small workshop with an audience of about 60 participants. The workshop is known for lively discussions and very active participation of the attendees. A significant component of the workshop is the institution of so-called Working Sessions chaired by distinguished researchers focusing on topical issues of the day. The Working Sessions, introduced in 1996 by Christophe Diot and Wallid Dabbous, have proved to be very successful, and they contribute considerably to making PfHSN a true "workshop."

This year, the program committee had to be once again rather selective, accepting only 14 out of 54 submissions as full papers. Working sessions on extremely timely issues, e.g., High-Speed Mobile Wireless, complemented the program. In addition, the workshop featured a keynote speech which gave an operator's viewpoint on high-speed networking, and an invited talk bringing a manufacturer's viewpoint. In honor of the large number of good submissions and to allow for the presentation of new and innovative work, the program was complemented by a set of six short papers and a panel session.

High-speed networking has changed enormously during the thirteen years covered by the workshop. Technologies such as ATM have moved into the spotlight and out again. What was once at the forefront of technology and deployed only in niches has become a commodity, with widespread availability of commercial products such as Gigabit Ethernet. At the same time, many issues identified by research to be important a decade ago have proven to be very timely today.

While this year's papers give answers to many important questions, they also show that there is still a lot of room for additional work in the future.

March 2002   Georg Carle, Martina Zitterbart

Organization

The Seventh International Workshop on Protocols for High-Speed Networks (PfHSN 2002), held in Berlin, Germany from Monday, April 22 to Wednesday, April 24, 2002, was jointly organized by the Fraunhofer Institute FOKUS and the Institute of Telematics, University of Karlsruhe. It was sponsored by IFIP WG6.2, the Working Group on Network and Internetwork Architecture. Technical co-sponsorship was provided by the IEEE Communications Society Technical Committee on Gigabit Networking. The workshop was organized in cooperation with COST Action 263 – Quality of future Internet Services.

Workshop Co-chairs

Georg Carle, Fraunhofer FOKUS, Germany
Martina Zitterbart, University of Karlsruhe, Germany

Steering Committee

James P.G. Sterbenz, BBN Technologies, GTE, USA (Chair)
Per Gunningberg, Uppsala University, Sweden
Byran Lyles, Sprint Labs, USA
Harry Rudin, IBM Zurich Research Lab, Switzerland
Martina Zitterbart, University of Karlsruhe, Germany

Program Committee

Sujata Banerjee, HP Labs and Univ. of Pittsburgh, USA
Olivier Bonaventure, University of Namur, Belgium
Torsten Braun, University of Bern, Switzerland
Georg Carle, Fraunhofer FOKUS, Germany
Jon Crowcroft, UCL, UK
Christophe Diot, Sprint Labs, USA
Julio Escobar, Centauri Technologies Corporation, Panama
Serge Fdida, University P. and M. Curie, Paris, France
Per Gunningberg, Uppsala University, Sweden
Marjory Johnson, RIACS/NASA Ames Research Center, USA
Guy Leduc, Univ. of Liege, Belgium
Jörg Liebeherr, University of Virginia, USA
Byran Lyles, Sprint Labs, USA
Gerald Neufeld, Redback Networks, USA
Luigi Rizzo, University of Pisa, Italy
Harry Rudin, IBM Zurich Research Lab, Switzerland

Table of Contents

Signalling and Controlling

A Core-Stateless Utility Based Rate Allocation Framework 1
Narayanan Venkitaraman, Jayanth P. Mysore, Mike Needham

Resource Management in Diffserv (RMD): A Functionality
and Performance Behavior Overview 17
*Lars Westberg, András Császár, Georgios Karagiannis, Ádám Marquetant,
David Partain, Octavian Pop, Vlora Rexhepi, Róbert Szabó,
Attila Takács*

Performance Evaluation of the Extensions
for Control Message Retransmissions in RSVP 35
Michael Menth, Rüdiger Martin

Application-Level Mechanisms

Handling Multiple Bottlenecks in Web Servers
Using Adaptive Inbound Controls 50
Thiemo Voigt, Per Gunningberg

Dynamic Right-Sizing: An Automated, Lightweight,
and Scalable Technique for Enhancing Grid Performance 69
Wu-chun Feng, Mike Fisk, Mark Gardner, Eric Weigle

The "Last-Copy" Approach for Distributed Cache Pruning
in a Cluster of HTTP Proxies 84
Reuven Cohen, Itai Dabran

TCP and High Speed Networks

Modeling Short-Lived TCP Connections
with Open Multiclass Queuing Networks 100
M. Garetto, R. Lo Cigno, M. Meo, E. Alessio, M. Ajmone Marsan

TCP over High Speed Variable Capacity Links:
A Simulation Study for Bandwidth Allocation 117
Henrik Abrahamsson, Olof Hagsand, Ian Marsh

TCP Westwood and Easy RED to Improve Fairness
in High-Speed Networks 130
    Luigi Alfredo Grieco, Saverio Mascolo

Quality of Service

A Simplified Guaranteed Service for the Internet 147
    Evgueni Ossipov, Gunnar Karlsson

Improvements to Core Stateless Fair Queueing 164
    Cristel Pelsser, Stefaan De Cnodder

A Fast Packet Classification by Using Enhanced Tuple Pruning 180
    Pi-Chung Wang, Chia-Tai Chan, Wei-Chun Tseng, Yaw-Chung Chen

Traffic Engineering and Mobility

Traffic Engineering with AIMD in MPLS Networks 192
    Jianping Wang, Stephen Patek, Haiyong Wang, Jörg Liebeherr

Performance Analysis of IP Micro-mobility Handoff Protocols 211
    Chris Blondia, Olga Casals, Peter De Cleyn, Gert Willems

Working Sessions

High-Speed Mobile and Wireless Networks 227
    James P.G. Sterbenz

Peer Networks – High-Speed Solution or Challenge? 228
    Joseph D. Touch

Invited Paper

High Speed Networks for Carriers 229
    Karl J. Schrodi

Protocols for High-Speed Networks:
A Brief Retrospective Survey of High-Speed Networking Research 243
    James P.G. Sterbenz

Author Index 267

A Core-Stateless Utility Based Rate Allocation Framework

Narayanan Venkitaraman, Jayanth P. Mysore, and Mike Needham

Networks and Infrastructure Research, Motorola Labs,
{venkitar,jayanth,needham}@labs.mot.com

Abstract. In this paper, we present a core-stateless framework for allocating bandwidth to flows based on their requirements which are expressed using utility functions. The framework inherently supports flows with adaptive resource requirements and intra-flow drop priorities. The edge routers implement a labeling algorithm which in effect embeds partial information from a flow's utility function in each packet. The core routers maintain no per-flow state. Forwarding decisions are based a packets label and on a threshold utility value that is dynamically computed. Thus the edge and core routers work in tandem to provide bandwidth allocations based on a flow's utility function. We show how the labeling algorithm can be tailored to provide different services like weighted fair rate allocations. We then show the performance of our approach using simulations.

1   Introduction

The Internet is being increasingly used for carrying multimedia streams that are sensitive to the end-to-end rate, delay and drop assurances they receive from the network. We are motivated by two key characteristics that a significant number of these flow share. First, multimedia flows are increasingly becoming adaptive and can adjust their level of performance based on the amount of resource available. The different levels of performance result in varying levels of satisfaction for the user. Another key characteristic is that most of them tend to be composed of packets which contribute varying amounts of utility to the flow they belong to. This intra-flow heterogeneity in packet utility could be caused due to the stream employing a hierarchical coding mechanism as in MPEG or layered multicast, or due to other reasons such as the specifics of a rate adaptation algorithm (as explained later for TCP). In either case, dropping the "wrong" packet(s) can significantly impact the qualitative and quantitative extent to which a flow is able to make use of the resources allocated to it. That being the case, the utility provided by a quantum of resource allocated to a flow depends on the value of the packets that use it. So, merely allocating a certain quantity of bandwidth to a flow does not always imply that the flow will be able to make optimal use of it at all times. As has been observed previously, applications don't care about

G. Carle and M. Zitterbart (Eds.): PfHSN 2002, LNCS 2334, pp. 1–16, 2002.

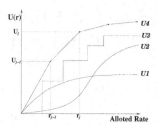

Fig. 1. Utility Functions

bandwidth, per se, except as a means to achieve user satisfaction. In summary, if optimizing the perceived quality of service is the end goal of an architecture, then it is important that we allocate resources according to user's preferences, and provide simple ways for a flow to make best use of its share.

Utility functions have long been recognized as an abstraction for a user to quantify the utility or satisfaction that (s)he derives when a flow is allocated a certain quantum of resource. It maps the range of operational points of a flow to the utility that a user derives at each point. Figure 1 shows some sample utility functions. Such an abstraction provides the necessary flexibility to express arbitrarily defined requirements. Also, it is now well established that different notions of fairness can be defined in terms of utility functions [4,7,8,10]. Partly motivated by recent work by Gibbens, Kelly and others [5,4,11,2], we use utility functions as an abstraction that is used to convey application/user level performance measures to the network. In this paper, we only concern ourselves with allocation of bandwidth as a resource. Therefore, we have used the terms resource and bandwidth interchangeably. We hope that the proposed framework will be a step toward a more general solution that can be used for allocation of other network resources such as those that impact end to end delay and jitter.

In this paper, we propose a scalable framework for allocating bandwidth to flows based on their utility functions. The architecture is characterized by its simplicity – only the edge routers maintain a limited amount of per flow state, and label the packets with some per-flow information. The forwarding behavior at a router is based on the state in the packet header. As the core routers do not perform flow classification and state management they can operate at very high speeds. Furthermore, the framework allows a flow to indicate the relative priority of packets within its stream. The dropping behavior of the system is such that for any flow lower priority packets are dropped preferentially over high priority packets of the flow.

We refer to our architecture as the *Stateless Utility based Resource allocation Framework*(SURF)[1]. The network objective and architecture are described in Section 2. The algorithms implemented by this architecture are described in Section 3. In Section 4, we present the performance of our architecture in a variety of scenarios. Section 5 discusses our implementation experience and some key issues. Section 6 discusses the related work and section 7 concludes the paper.

---

[1] We borrow the notion of stateless core from CSFQ [14]

2 System Architecture

2.1 Network Model

The approach that we propose is based on the same philosophy used in technologies like CSFQ [14] and Corelite [12]. The network's edge routers maintain per-flow state information, and label packets based on rate at which flows send packets. Core routers maintain no per-flow state. Forwarding decisions are based on the labels that packet carry and aggregate state information such as queue length. Thus the edge and the core routers work in tandem to provide per-flow allocations. We build on these principles to provide resource allocation based on utility functions.

2.2 Network Objective

Let us assume that all flows provide the network their utility functions. There are a variety of objective functions that can be used to accomplish different goals. In the following discussion we consider two possible objectives, provide the intuition behind them, and motivate our choice of one of them as an objective that we use in this paper.

A possible network objective is to maximize the aggregate utility at every link in the network. i.e., at every link in the network, maximize $\sum_{i=1}^{M} U_i(r_i)$, subject to the constraint $\sum_{i=1}^{M} r_i \leq C$, where M is the number of flows sharing the link, $U_i(r_i)$ is the utility derived by flow i for a allocation r_i and C is the total link capacity. Another potential objective is to maximize the aggregate system utility, i.e, maximize $\sum_{i=1}^{N} U_i(r_i)$, where N is the total number of flows in the network. For a network with just a single link both the objectives are identical. However, for a multi-hop network they are different. For instance, consider the example shown in Figure 2. Here, $f1$ is a high priority flow and hence has a larger incremental utility than that for $f2$ and $f3$. If the available bandwidth is two units, then if we use the first objective function we will allocate both units to $f1$ in both the links. This maximizes the utility at every link in the network(3.0 units at every link) and the resultant system utility is 3.0 because only $f1$ received a bandwidth allocation. However, if we use the second objective function, we will allocate two units to $f2$ and $f3$ in each of the links. Though the aggregate utility at any given link is only 2.0, the resultant aggregate system utility is 4.0. This difference in allocation results from different interpretations

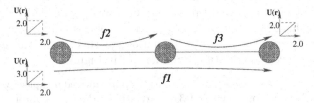

Fig. 2. Bandwidth Allocation Example

of the utility function. The first objective function treats utility functions from a user's perspective by collapsing the entire network into a single unified resource, neglecting the hop count. Thus, this interpretation has 2 key characteristics: (i) it is topology agnostic, i.e. a user does not have to be concerned with how many hops a flow traverses when specifying a utility value, and (ii) it maintains the relative importance of different flows as specified by the utility function across all links.

While the first objective function suites an user's perspective the second treats them from a resource pricing point of view. In this interpretation, the more hops a flow traverses the more resources the flow utilizes and the more a user should pay for comparable performance. Specifically, the utility functions can be viewed as quantifying a user's willingness to pay. Optimizing the second objective function can maximize the network operator's revenue. Networks which employ such an optimization criterion require a user to be cognizant of the hop count of the end to end path traversed by his flow and alter the utility function to get performance comparable to a case with a different number of hops.

Arguably, a case can be made in favor of either of the cases mentioned here or many other possible objectives. As our focus in this paper is to view utility function as a guide to user satisfaction, independent of network topology, we choose to focus on the former objective function.

3   Distributed Framework and Algorithms

In this section, we describe the distributed framework that provides rate allocations that approximate the desired network objective. A key characteristic of the framework is that only the routers at the edges of the network maintain per-flow state information and have access to the utility function of the flows. The core routers however, treat packets independent of each other. They do not perform any per-flow processing and have a simple forwarding behavior.

The framework has two primary concepts: First, an ingress edge router logically partitions a flow into *substreams*. The substreams correspond to different slopes in the utility function of the flow. Substreaming is done by appropriately labeling the headers of packets using incremental utilities. Second, a core router has no notion of a flow, and treats packets independent of each other. The forwarding decision at any router is solely based on the incremental utility labels on the packet headers. Routers do not drop a packet with a higher incremental utility label as long as a lower priority packet can instead be dropped. In other words, the core router attempts to provide the same forwarding behavior of a switch implementing a multi-priority queue by using instead a simple FIFO scheduling mechanism, eliminating any need for maintaining multiple queues or sorting the queue. For ease of explanation, in this paper, we describe the algorithms in the context of utility function U4 in Figure 1[2].

---

[2] Many utility functions such as U1 can be easily approximated to a piece-wise function similar U4. For functions such as U3 we are still working on appropriate labeling algorithms that provide the right allocation with least amount of oscillations

3.1   Substreaming at the Edge

Every ingress edge router maintains the utility function, $U(r)$, and the current sending rate, r, corresponding to every flow that it serves. The current sending rate of a flow can be estimated using an algorithm similar to the one described in CSFQ [14]. The edge router then uses a labeling algorithm to compute an incremental utility value, u_i, that should be marked on the packet header. The result of this procedure is that the flow is logically divided into k substreams of different incremental utilities, where k is the number of regions or steps[3] in the utility function from 0 to r. The u_i field is set to $(U(r_j) - U(r_{j-1}))/(r_j - r_{j-1})$ which represents the increment in utility that a flow derives per unit of bandwidth allocated to it, in the range (r_j, r_{j-1}). Thus all packets have a small piece of information based on the utility function of the flow embedded in them.

3.2   Maximizing Aggregate Utility

Routers accept packets such that a packet with a higher incremental utility value is not dropped as long as a packet with a lower incremental utility could instead be dropped. Such a policy ensures that in any given router, the sum of u_i of the accepted packets is maximized. There are many different ways by which such a dropping policy can be implemented in the router.

One solution is to maintain a queue in the decreasing order of priorities[4]. When the queue size reaches its maximum limit, q_{lim}, the lowest priority packet in the queue can readily be dropped and incoming packet can be inserted appropriately. This would provide the ideal result. But in a high speed router, even with a moderate queue size, such a solution will be inefficient as the processing time allowable for any given packet will be very small. In the following section, we propose an algorithm that approximates the behavior of such a dropping discipline using a simple FIFO queue, without the requirements of maintaining packets in a sorted order or managing per-flow or per-class information.

Priority Dropping with a FIFO Queue: The problem of dropping packets with lower incremental utility labels before packets with a higher incremental utility can be approximated to the problem of dynamically computing a minimum threshold value that a packet's label must have, in order for a router to forward it. We call this value the threshold utility, u_t. We define threshold utility as the minimum incremental utility that a packet must have for it to be accepted by the router. The two key constraints on u_t are that it must be maintained at a value which will (a) result in enough packets being accepted to fully utilize the link and (b) not cause buffer overflow at the router.

In Figure 3, $R(u)$ is a monotonically decreasing function of the incremental utility u. It represents the cumulative rate of all packets that are forwarded

---

[3] A step in refers to a contiguous region of resource values with the same slope

[4] This will be in addition to the FIFO queue, that is required to avoid any reordering of packets.

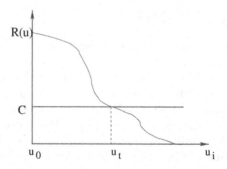

Fig. 3. Threshold Utility

through a link for a given threshold utility value, u_j. So $R(u_j) = \sum_{k=j}^{max} r(u_k)$, where $r(u_k)$ is the rate of packets entering an output link with an incremental utility label of u_k. The threshold utility, u_t is a value which satisfies the condition $R(u_t) = C$, where C is the capacity of the output link. Note that for a given $R(u)$, there may not exist a solution to $R(u) = C$ because of discontinuities in $R(u)$. Also the function $R(u)$ changes with time as flows with different utility functions enter and leave the system and when existing flows change their sending rates. Hence, tracking the function is not only expensive but may in fact be impossible. So in theory, an algorithm that uses the value of a threshold utility for making accept or drop decisions, cannot hope to replicate the result obtained by an approach that involves sorting using per-flow state information. Our objective is to obtain a reasonably close approximation of the threshold utility so that the sum of utilities of flows serviced by the link closely tracks the optimal value, while the capacity of the output link is fully utilized.

First, we give the intuition behind the algorithm that a router uses to maintain the threshold utility u_t for an output link and then provide the pseudo code for the algorithm. We then describe how it is used to make the forward or drop decision on a new incoming packet.

The objectives of the algorithm are (i) to maintain the value of u_t such that for the given link capacity the sum of the utilities of all the accepted packets is close to the maximum value possible, and (ii) to maintain the queue length around a specified lower and upper threshold values(q_{lth} and q_{uth}). There are three key components in the algorithm. (a) to decide whether to increase, decrease or maintain the current value of u_t, (b) to compute the quantum of change and (c) to decide how often u_t should be changed. The key factors that determine these decisions are avg_qlen, an average value of the queue length computed (well known methods for computing the average, like the exponential averaging technique can be used for this purpose) and q_{dif}, the difference between the virtual queue length value at the current time and when the threshold u_t was last updated. The virtual queue length is a value that is increased on an enqueue event by the size of the packet received if its $u_i \geq u_t$. The value is decreased by the size of the packet either on a deque or during a successful enque of a packet with a label less than than the $u_t{}^5$. The latter enables corrective action when packets are being dropped due to a incorrect(large) threshold. Thus, the virtual queue length is simply a value that increases and decreases without the physical constraints of a real queue. Maintaining a virtual queue length in this manner provides an accurate estimate of the state of congestion in the system. Note that even when the real queue overflows, the virtual queue length will increase, resulting in a positive q_{dif} reflecting the level of congestion. Similarly, when the u_t value is very large and no packets are being accepted, the virtual queue length will decrease, resulting in a negative value of q_{dif}. q_{dif} reflects the rate at which the length of the virtual queue is changing. When there is a sudden change in $R(u)$, q_{dif} provides a early warning signal which indicates that u_t may need to be modified. However, if the link state changes from uncongested to congested slowly, the absolute value of q_{dif} may remain small. But a value of avg_qlen that is beyond the specified queue thresholds indicates that u_t needs to be changed.

The quantum of change applied to u_t is based on the amount of buffer space left – given by the queue length, and the rate at which the system is changing – given by q_{dif}. Congestion build up, is equivalent to $R(u)$ in Fig. 3 shifting to the right. To increase the threshold, we use a heuristic to determine a target value u_{itgt} such that $R(u_{itgt}) < C$. This is used to significantly reduce the probability of tail drops. Currently this value of u_{itgt} is based on the average u_i values of all the accepted packets and the maximum u_i value seen in the last epoch. The value of u_t is then incremented in step sizes that are based on the estimated amount of time left before the buffer overflows. Similar computation is done to decrease the threshold where u_{dtgt} is based on the average u_i values of all packets dropped in the last epoch. The pseudocode for updating the threshold is as follows:

$$
\begin{aligned}
&\text{if } (avg\_qlen < q_{lth})or(q_{dif} < -K_q)\\
&\qquad time\_left = avg\_qlen/q_{dif}\\
&\qquad change = (u_t - u_{dtgt})/time_left\\
&\text{else if } (avg\_qlen > q_{uth})or(q_{dif} > K_q)\\
&\qquad time\_left = (q_{lim} - avg\_qlen)/q_{dif}\\
&\qquad change = (u_{itgt} - u_t)/time\_left\\
&u\_t+ = change
\end{aligned}
$$

There are two events that trigger a call to the update-threshold() function. They are (a) whenever $|q_{dif}| > K_q$ and (b) a periodic call at the end of a fixed size epoch. The first trigger ensures fast reaction. The value of K_q is a configurable parameter and is set such that we do not misinterpret a typical packet burst as congestion. Also, it provides a self-healing feedback loop. For instance, when congestion is receding, if we decrease u_t by steps that are smaller than optimal, this trigger will result in the change being applied more often. Case (b) ensures that during steady state, the value of u_t is adjusted so that the queue length is maintained within the specified queue thresholds.

---

[5] This case would occur when link is not congested but u_t is incorrectly large.

The forwarding algorithm is very simple. When the link is in a congested state($avg\_qlen > q_{lth}$), if $u_i \le u_t$ the packet is dropped. Otherwise the packet is accepted.

3.3   Variants of SURF

The framework described above is flexible and can be tailored to specific needs by choosing appropriate utility functions. The labeling algorithms at the edge router can be tailored to label the incremental utilities for specific cases. The forwarding and threshold computation algorithms remain the same. This is a big advantage. In this section, we describe a few specific cases of the edge labeling algorithm and provide the pseudo-code.

Fair Bandwidth Allocation. A common notion of fair bandwidth allocation is one in which all flows that are bottle necked at a link get the same rate, called the fair share rate. To achieve such an allocation, all we need to do is assign identical utility functions with constantly decreasing incremental utilities to all flows. For ease of understanding we provide a labeling procedure for an idealized bit-by-bit fluid model. Let u_{max} be the maximum possible value of incremental utility.

label(pkt)
   $served+ = 1$
   $pkt.u_i = u_{max} - served$

where the value of *served* is reset to 0, after a fixed size labeling epoch, say 1 sec. Let us suppose that the rate at which each flow is sending bits is constant. The result of this labeling algorithm then is that during any given second, the bits from a flow sending at rate r bits per second are marked sequentially from u_{max} to $u_{max} - r$. The router in a bottleneck link will compute the threshold u_t and drop packets from all flows with $u_i < u_t$. This results in fair bandwidth allocation. This is an alternate implementation of CSFQ [14]. As we will see in the next section, a key advantage of this approach is that it allows us to convey rate information as well as intra-flow utility using the same field in the packet header.

Intra-flow Priorities. Consider a flow, i, which is sending packets with multiple priority levels at a cumulative rate r_i. For instance, the levels could be I, P and B frames in an MPEG video stream or layers in a layered multicast stream [9]. The utility function corresponding to the flow will be similar to U4 in Figure 1. Independent of the number of layers and rate allocated to other flows, if flow i's packets need to dropped, we would like the packets from layer $j + 1$ to be dropped before layer j. To achieve such a dropping behavior, the end hosts must communicate the relative priority of a packet to the edge router. A simple mechanism to accomplish this could be in the form of a field in the packet header. The desired dropping behavior honoring intra-flow drop priorities can be achieved in the proposed framework by using a labeling procedure similar to the pseudo-code given below. The forwarding and threshold computation algorithms remain unchanged.

```
label(pkt)
    p = pkt.intraflow_priority
    served[p]+ = pkt.size
    cum_rate[p] = cum_rate[p − 1] + est_rate[p]
    if (served[p] < cum_rate[p − 1]) or
        (served[p] ≥ cum_rate[p])
            served[p] = cum_rate[p − 1]
    pkt.uᵢ = u(served[p])
```

Figure 4 describes the above pseudo-code. In the code given above, $est\_rate[p]$ is the estimated rate at which a flow is sending packets of a certain priority level p and $cum\_rate[p]$ is simply $\sum_{i=1}^{p} est\_rate[p]$ (assuming 1 to be the highest priority level). $est\_rate[p]$ can be computed using a method similar to the one used in [14]. $served[p]$ maps the packet received onto the appropriate region in its utility function. $u(r)$ gives the incremental utility of the region corresponding to rate r.

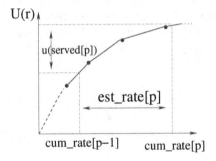

Fig. 4. Labeling Algorithm

Improving TCP Performance. The labeling algorithm used by the edge router can be tailored to improve the performance of TCP. Specifically, it can mitigate two primary causes of a reduction in throughput – (i) drop of a retransmitted packet (ii) back to back drops of multiple packets. To accomplish the former, the labeling procedure can label retransmitted packets with the highest allowable incremental utility for the flow; and to accomplish the latter, the labeling algorithm can assign consecutive packets to different priority levels thereby reducing the chances of back-to-back drops. Such an implicit assignment of interleaved priorities is also useful for audio streams, whose perceived quality improves when back-to-back drops are avoided.

4   Performance Evaluation

In this section, we evaluate the algorithms using simulations. We have implemented the algorithms in the *ns-2* simulator. We have used the two topologies. Topology 1 is a simple network with a single congested link which all flows share. We use this topology to first show that the system converges to the optimal rate allocation with both adaptive and non adaptive flows. We then use topology 2, shown in Figure 5, to confirm that the solution converges to the expected values in a multi hop network. We then take a specific variant of SURF discussed in 3.3, where all flows have the same utility function. Using TCP and CBR flows we show that fair allocation is achieved. Finally, using a flow with 3 different packet priority levels, we show that the proposed framework drops packets based on a flow's intra-flow priority. The allocated rate for CBR flows is measured by summing the number of bits received every 250 ms in the sinks, and that for adaptive flows is measured in the sinks using an exponential averaging routine(similar to the arrival rate estimation algorithm in the edge router).

All the simulations presented in this section use a fixed packet size of 1000 bytes, a maximum queue size of 100 packets, queue thresholds q_{uth} and q_{lth} of 10 and 50 packets respectively and a K_q of 10 packets. K_q and q_{lth} are chosen so that we do not interpret small bursts that typically occur at a router as congestion and q_{uth} is chosen so that there is sufficient time to adapt u_t and avoid tail drops. Flows use the utility functions shown in Figure 6. The 4 values given for each utility function are incremental utilities for the 4 rate regions. For instance, function U2 has an incremental utility of 0.45 for an allocation in the region 480-960Kbps.

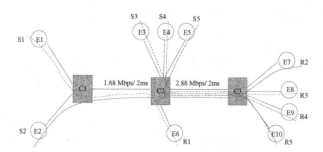

Fig. 5. Topology 2

4.1   Single Congested Link

In this scenario, Flows 1, 2 and 3 have utility functions $U1$, $U2$ and $U3$ (shown in Figure 6) respectively. For the first set of results, we consider 3 non-adaptive CBR sources, each sending at a rate of 1.92Mbps, the link capacity. To calculate the optimal allocations, we note that the incremental utilities for these flows are

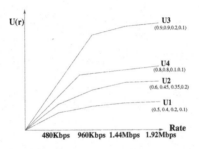

Fig. 6. Utility Functions

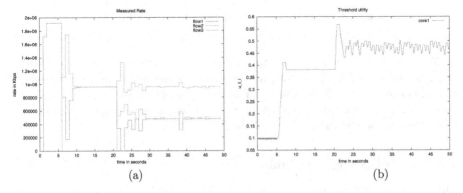

Fig. 7. Single Link With Non Adaptive Flows

such that, $U3$ has the highest incremental utility in region 0-960Kbps. This is followed by $U2$ then $U1$ in the region 0-480Kbps, followed by $U2$ and $U1$ again in the region 480-1.44Mbps. So when flow1 and flow2 are sharing the channel, the optimal allocation would be to provide 960Kbps each to flow1 and flow2. Similarly when all 3 flows share the channel, the optimal allocation is to allocate 960Kbps to flow3 and 480Kbps each to flow1 and flow2.

In the simulation, flow 1, enters the network at time 0. As it is the only flow in the system, it is allocated the entire network bandwidth. 5 seconds after the start of the simulation, flow 2 enters the network. From Figure 7(a) we see that flow1 and flow2 are allocated and 960Kbps each, which is the optimal allocation. After 20 seconds, flow 3, enters the system. This flow has incremental utility of 0.9 (U3 of Figure 6) which is larger than that of the other two flows in the region 0-960Kbps. Consequently, the threshold utility, shown in Figure 7(b), momentarily shoots up. However, this results in increased packet drops, resulting in containing the queue size within acceptable limits. When the queue size drops, the threshold utility also drops to a lower value. It finally stabilizes at a value that results in the capacity being shared between Flows 1, 2 and 3 in the ratio 1:1:2. This conforms to the allocation that optimizes the aggregate utility of the flows sharing the link.

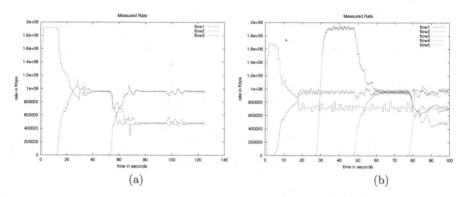

Fig. 8. Adaptive and non-adaptive flows with SURF

We now present the results for the same scenario with adaptive flows. Flow 2 and 3 are adaptive. As these flows adjust their sending rates, the distribution of u_i (and hence, $R(u)$) will also change. The objective of this results is to show that even when $R(u)$ changes u_t converges to the correct value. Figure 8(a) shows the measured rate at the receivers of the flows.

4.2   Multiple Congested Links

The results for this scenario are shown in Figure 8(b). In this case, there are five flows having utility functions U1,U2,U3,U1 and U4 respectively. The key objectives in this experiment are to show (1) that the computation of threshold utility settles about the correct value even with flows traversing more than just a single link, (2) even if there is no exact solution to $R(u) = C$, the value computed by the algorithm hovers around the right value.

Between time 0 to 25 seconds, only flow 1 and 2 are in the system. The optimal allocation will result in a ratio of 2:1.5 for flows 1 and 2. This is a case when there is no exact solution to the condition $R(u) = C$. At $u_t < 0.4$, the accepted rate will be only 1.44Mbps, where as at $u_t \geq 0.4$, the accepted rate will be 1.92Mbps. The results in Figures 8(b) show that the bandwidth distribution stabilizes around the correct values.

At time 25 seconds, flow 3 enters and occupies the entire remaining bandwidth in the link C2-C3. Its entry does not affect the allocations made to flow2. Flow4 then enters at time 45sec. It grabs its share of the two units of bandwidth at the expense of flow3 because flow 3 has a lower incremental utility in the region beyond 960Kbps, compared to flow 4 in the region 0-960Kbps. At time 75sec, flow 5 enters the network. It has a larger incremental utility till the rate 720Kbps, compared to flows 2 and 4. So it gets that share in preference to flows 2 and 4, which end up with allocations of 720Kbps and 480Kbps respectively.

4.3   Achieving Fair Allocation

The simulated network topology has a single link with 20 sources sharing a 10Mbps. The labeling epoch length in the edge routers is set to 250 ms. The edge routers implement the labeling algorithms that provide fair bandwidth allocation. The sources successively enter the system in 200 ms intervals. Figure 9(a) and (b) show the rate allocations and the profile of the queue length with TCP sources. The figures indicate that: (1) there is approximately fair sharing of the link bandwidth, (2) the queue length is controlled to a range close to the specified threshold values.

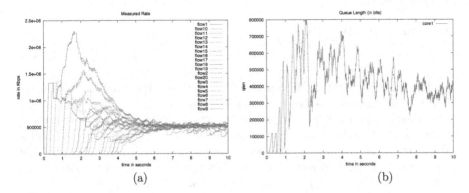

(a)   (b)

Fig. 9. Providing Fair Allocation For TCP Flows

4.4   Honoring Intra-flow Priorities

We next perform a simple experiment to illustrate the support for intra-flow priority dropping provided by an island of SURF routers. In this experiment we have five UDP flows sharing a 2 Mbps link. Each of the flows is assumed to be encoded using a layered encoding scheme that has 3 levels of priority. A base layer of highest priority(priority 1), and two enhancement layers with priorities 2 and 3(priority 2 packets being more useful in reconstructing the content than those of priority 3). Further, the flows send a third of their packets in each layer, uniformly striating the flow across the three layers. Each flow sends data at a rate of 1 Mbps, resulting in a fair share of 400 Kbps. When using SURF, we observed that each flow was in fact allocated exactly 400 Kbps.

$Priority$	$total\_sent$	$dropped$
1	785	3
2	785	618
3	785	761

The above table shows the dropping behavior of a network implementing SURF. The values shown are typical of any given flow. $total\_sent$ is the number of packet sent by a flow. *dropped*, the number of packets dropped in different priority levels clearly indicates that the lower priority packets are dropped in preference to higher priority packets. During the initial phase of the simulation, when flows are introduced successively, all the layers are served. But as congestion builds up, all packets from priority 3 are dropped. The fair share of 400Kbps is shared between the entire 333Kbps of priority 1 packets and a small portion(66Kbps) of priority 2 packets. This simulation shows that in addition to providing utility based allocation of rate, SURF routers honor intra-flow priorities.

5   Discussion

5.1   Implementation Experience

We have implemented the framework described here in the linux kernel, as a queuing discipline that can be configured using *'tc'*, the traffic control program. The threshold computation, forwarding procedure and the different variants of SURF that we discussed before have been implemented. The results obtained using the simulations match closely with the results on our implementation testbed. We have used *vcr*, a client-server program that can stream mpeg audio-video files and play the same, to demonstrate the improvement in user experience. The only change that was required to the *vcr* program itself, was an addition to convey the relative priority of a packet. We use the standard *'setsockopt*()' system call before every *'send*()' operation to convey this information to the IP layer. The time and space complexity of implementing the operations of the core router are of the same order as RED. So we believe that the proposed schemes can be implemented without compromising the efficiency of high speed core routers.

5.2   Convergence Issues

One of the key issues is the stability and convergence of the threshold computation algorithm. Though in the simulations that we have performed the threshold converges to the correct value, an analytical evaluation of the convergence of the algorithm is essential. In [1], we have considered the specific case of section 3.3 where all flows have the same utility function. We consider the different cases with constant bit rate sources and TCP like rate adaptive sources and on-off sources. We have derived the conditions that are essential for the convergence of a threshold computation algorithm(or fair share estimation in the CSFQ). We show that both the CSFQ algorithm and the algorithm proposed in this paper satisfy those conditions. Further analysis with different types of utility function is part of ongoing work.

5.3   Pricing Issues

In a network that provides service differentiation based on utility functions, greedy users may be tempted to associate the highest utility value to any amount of bandwidth allocated. Typically a pricing scheme is used to prevent users from lying about their utility functions. The proposed framework does not dictate the use of a specific pricing scheme. This framework can coexist with many different pricing schemes. In many cases, the pricing scheme can use some parameters provided by the framework to determine the cost of the resource. For example, in a simple spot-market pricing mechanism that is based on the current level of congestion, the threshold utility(u_t) can be used as an indicator of the current the level of demand of the resource. Also, note that the framework does not bind a flow to a particular utility function. It offers the flexibility to dynamically change the utility function during the lifetime of a flow. This is a very useful feature because, the utility that a user derives from a certain amount of resource allocated to a flow can change dynamically based on various circumstances.

6   Related Work

The concept of core-stateless networks was proposed by Stoica et al in [14]. They provide an architecture for approximating the behavior of a network of fair queuing routers. [14] achieves this by labeling packets with the estimated sending rate of a flow and probabilistically dropping packets based on the label and a dynamically computed fair share rate in the core. Our approach is partly motivated by their work. In [13], the authors propose a similar approach based on carrying flow state in packet headers to provide delay guarantees. RFQ [3], which was developed independent of our framework is an extension to CSFQ which provides support for intra-flow priorities by marking packet with different colors. It does not consider user-specified utility functions for labeling packets. In [11], Kunniyur and Srikant, describe a framework for designing end-to-end congestion control schemes where users can have different utility functions. While the framework proposed in this paper is network based, the framework in [11] depends on flows performing congestion control based on utility functions. The objective function they optimize is different from the one we elaborate in this paper. In fact, they maximize aggregate user satisfaction, rather than utility at every link. Also, being a purely end-to-end adaptation scheme with no router support, it does not provide for intra-flow drop priorities. [6] presents a framework for optimizing the aggregate user satisfaction of a network by performing congestion avoidance using the congestion price as the feedback parameter. They present an algorithm core routers can use for computing the congestion price (the shadow price), and a rate adaptation algorithm for end hosts. The rate adaptation algorithm adapts the rate of a flow based on a flows willingness to pay. They show that at the socially optimum allocation, the first derivative of a user's utility function exactly matches the sum of the shadow prices of all resources along the flows route. They propose changes to the use of the ECN mechanism in TCP, to implement their framework. [2] proposes a mechanism of utility max-min which tries to maximize the minimum utility received by applications sharing a bottleneck link. This is different from the objectives that we considered in section 2.

7   Conclusion

In this paper, we have presented *SURF*, a scalable architecture for providing bandwidth allocation to flows based on their utility functions. Our proposed architecture maximizes the aggregate utility at each link. Its key attributes are scalability, accomplished using a core-stateless architecture, and support for intra-flow priorities. By simply tailoring edge labeling algorithms, this framework can be leveraged to optimize the performance of a variety of flows. We described an algorithm for computing the threshold utility and forwarding packets based on the computed value. We then presented labeling algorithms that can be used to provide fair sharing of link bandwidth; and to optimize the perceived quality of layered streams while providing equal sharing of the bandwidth across flows. We presented a selected set of results to illustrate various facets of performance. As high speed networks become increasingly common, we believe that core-stateless schemes such as SURF will be useful in enabling flexible service models and optimizing user satisfaction in a unified, scalable framework.

References

1. R. Barnes, R. Srikant, J.P. Mysore, N. Venkitaraman "Analysis of Stateless Fair Queuing Algorithms", *to appear in Proc. of 35th Annual Conference on Information Sciences and Systems*, March 2001.
2. Z. Cao and E. Zegura. "Utility Max-Min: An application-Oriented Bandwidth Allocation Scheme", *Proc. of IEEE INFOCOM*, 1999.
3. Z. Cao, Z. Wang and E. Zegura. "Rainbow Fair Queuing: Fair Bandwidth Sharing Without Per-Flow State", *Proc. of IEEE INFOCOM*, 2000.
4. F.P. Kelly. "Charging and Rate Control for Elastic Traffic", *European Transactions on Telecommunications*, vol 8, 1997.
5. R.J. Gibbens and F.P. Kelly. "Distributed connection acceptance control for a connectionless network", *Proceedings of the 16th International Teletraffic Congress, Edinburg*, June 1999.
6. P. Key, D. McAuley, P. Barham. "Congestion Pricing for Congestion Avoidance", *Technical Report MSR-TR-99-15, Microsoft Research*, Febuary 1999.
7. L. Massoulie and J. Roberts. "Bandwidth Sharing: Objectives and Algorithms", *Proc. of INFOCOM*, March 1999.
8. J. Mo and J. Warlang. "Fair end-to-end Window based Congestion Control",
9. S. McCanne, V. Jacobson, and M. Vetterli. "Receiver-driven layered multicast", *Proc. SIGCOMM'96*, Stanford, CA, Aug. 1996, ACM, pp. 117-130.
10. S. Shenker "Fundamental Design Issues for the Future Internet", *IEEE JSAC. Vol 13, No. 7*, September 1995.
11. S. Kunniyur and R. Srikant "End-to-End Congestion Control Schemes: Utility Functions, Random Losses and ECN Marks", *Proc. of IEEE INFOCOM* 2000.
12. S. Raghupathy, T. Kim, N. Venkitaraman, and V. Bharghavan. "Providing a Flexible Service Model with a Stateless Core", *Proc. of ICDCS 2000*.
13. I. Stoica and H. Zhang. "Providing Guaranteed Services Without Per Flow Management", *Proceedings of the ACM SIGCOMM '99 Conference*, September 1999.
14. I. Stoica, S. Shenker and H. Zhang. " Core-Stateless Fair Queueing: Achieveing approximately Fair Bandwidth Allocations in High Speed Networks", *Proceedings of the ACM SIGCOMM '98 Conference*, September 1998.

Resource Management in Diffserv (RMD):
A Functionality and Performance Behavior Overview

Lars Westberg[1], András Császár[2], Georgios Karagiannis[3], Ádám Marquetant[2],
David Partain[4], Octavian Pop[2], Vlora Rexhepi[3], Róbert Szabó[2,5], and Attila Takács[2]

[1] Ericsson Research
Torshamnsgatan 23 SE-164 80, Stockholm, Sweden
lars.westberg@era-t.ericsson.se
[2] Net Lab, Ericsson Research Hungary
Laborc u. 1., Budapest H-1037, Hungary
{robert.szabo,andras.csaszar,adam.marquetant,octavian.pop,
attila.takacs}@eth.ericsson.se
[3] Ericsson EuroLab Netherlands,
P.O. Box 645, 7500 AP Enschede, The Netherlands
{vlora.rexhepi,georgios.karagiannis}@eln.ericsson.se
[4] Ericsson Radio Systems AB
P.O. Box 1248 SE-581 12 Linkoping, Sweden
david.partain@ericsson.com
[5]High Speed Networks Laboratory, Department of Telecommunications and Telematics,
Budapest University of Technology and Economics
Stoczek u. 2, Budapest H-1111, Hungary
robert.szabo@ttt.bme.hu

Abstract. The flexibility and the wide deployment of IP technologies have driven the development of IP-based solutions for wireless networks, like IP-based Radio Access Networks (RAN). These networks have different characteristics when compared to traditional IP networks, imposing very strict requirements on Quality of Service (QoS) solutions, such as fast dynamic resource reservation, simplicity, scalability, low cost, severe congestion handling and easy implementation. A new QoS framework, called Resource Management in Differentiated Services (RMD), aims to satisfy these requirements. RMD has been introduced in recent publications. It extends the IETF Differentiated Services (Diffserv) architecture with new admission control and resource reservation concepts in a scalable way. This paper gives an overview of the RMD functionality and its performance behavior. Furthermore, it shows that the mean processing delay of RMD signaling reservation messages is more than 1330 times smaller then the mean processing delay of RSVP signaling reservation messages.

Introduction

Internet QoS has been one of the most challenging topics of networking research for several years now. This is due to the diversity of current Internet applications, ranging from simple ones like e-mail and World Wide Web (WWW) up to demanding real-

G. Carle and M. Zitterbart (Eds.): PfHSN 2002, LNCS 2334, pp. 17–34, 2002.

time applications, like IP telephony, which are increasing the demand for better performance on the Internet.

Currently, due to flexibility and wide deployment of IP technologies, IP-based solutions have been proposed for wireless networks, like IP-based Radio Access Networks (RAN). These networks have different characteristics when compared to traditional IP networks, imposing very strict requirements on QoS solutions [19]. These QoS requirements are fast dynamic resource reservation, simplicity, low costs, severe congestion handling and easy implementation along with good scalability properties.

The Internet Engineering Task Force (IETF) standardization body is starting a new Working Group (WG), called Next Steps In Signaling (NSIS) [5] to specify and develop new types of QoS signaling solutions that will meet the real time application requirements and the QoS requirements imposed by the IP-based wireless networks. Several resource reservation mechanisms defined in the context of IP networks might be used as input to this WG. The most promising are RSVP (Resource reSerVation Protocol) [6], RSVP aggregation [7], Boomerang [8], YESSIR (YEt another Sender Session Internet Reservation) [9], Feedback control extension to differentiated services [10], Dynamic packet states [11], and Dynamic Reservation Protocol (DRP) [12].

While these schemes are able to satisfy the expectations that users of demanding real-time applications have, they fail to meet the strict QoS requirements imposed by IP-based wireless networks.

This paper presents a new QoS framework, called Resource Management in Differentiated Services (RMD), which aims to correct this situation. RMD is introduced in [13], [14], [15], [16], [17] and [18], and it represents a QoS framework that extends the Diffserv architecture with new admission control and resource reservation concepts in a scalable way. Even though it is optimized for networks with fast and highly dynamic resource reservation requirements, such as IP-based cellular radio access networks [19], it can be applied in any type of Diffserv networks. This paper presents a general overview of the RMD concept, functionality and its performance behavior.

The organization of this paper is as follows: Section 2 gives an introduction of the RMD fundamental concepts. The two RMD operation types, i.e., normal operation and fault handling operation are described in Section 3. Section 4 describes the performance evaluation experiments that were used to observe the performance behavior of the RMD framework. Finally, the conclusion and future work is given in Section 5.

RMD Fundamental Concepts

The RMD proposal described in this paper is based on standardized Diffserv principles for traffic differentiation and as such it is a single domain, edge-to-edge resource management scheme. RMD extends the Diffserv principles with new ones necessary to provide dynamic resource management and admission control in Diffserv domains. In general, RMD is a rather simple scheme.

Basic Idea behind RMD

The development of RMD was initially based on two main design goals. The first one was that RMD should be stateless on some nodes (e.g. interior routers), i.e., no per-flow state is used, unlike RSVP [6] , which installs one state for each flow on all the nodes in the communication path. The second goal was that RMD even though stateless should associate each reservation to each flow and therefore should provide certain QoS guarantees for each flow.

These two goals are met by separating a complex reservation mechanism used in some nodes from a much simpler reservation mechanism needed in other nodes.

In particular, it is assumed that some nodes will support "per-flow states", i.e., are stateful. In RMD these nodes are denoted as "edge nodes". However, any nodes that maintain reservation states could fulfill this requirement.

The second assumption is that the nodes between these stateful nodes can have a simpler execution by using only one aggregated reservation state per traffic class. In RMD these nodes are denoted as interior nodes.

The edges will generate reservation requests for each flow, similar to RSVP, but in order to achieve simplicity in interior nodes, a measurement-based approach on the number of the requested resources per traffic class is applied. In practice, this means that the aggregated reservation state per traffic class in the interior nodes is updated by a measurement-based algorithm that uses the requested and available resources as input. Unlike typical measurement based admission control (MBAC) algorithms, that apply admission control using data traffic measurements and available resources as input, RMD applies admission control on resource parameter values included in the reservation requests, i.e. signaling messages and available resources per traffic class.

RMD Protocols

The scalability of the Diffserv architecture is achieved by offering services on an aggregated basis rather than per flow and by forcing the per-flow state as much as possible to the edges of the network. The Differentiated Services (DS) field in the IP header and the Per-Hop Behavior (PHB) are the main building blocks used for service differentiation. Packets are handled at each node according to the PHB indicated by the DS field (i.e., Differentiated Service Code Point (DSCP) [4]) in the message header. However, the Diffserv architecture currently does not have a standardized solution for dynamic resource reservation.

The RMD framework is a dynamic resource management developed on Diffserv principles, which does not affect its scalability. This is achieved by separation of the complex per domain reservation mechanism from the simple reservation mechanism needed for a node. Accordingly, in the RMD framework, there are two types of protocols defined: the Per Domain Reservation (PDR) protocol and the Per Hop Reservation (PHR) protocol. The PDR protocol is used for resource management in the whole Diffserv domain, while the PHR protocol is used for managing resources for each node, on per hop basis, i.e., per DSCP. The PDR protocol can either be a newly defined protocol or an existing one such as RSVP [6] or RSVP aggregation [7], while the PHR protocol is a newly defined protocol. So far there is only one PHR protocol specified, the RMD On-demand (RODA) PHR [14] protocol.

This is made possible by definition of the Per Domain Reservation (PDR) protocol and the Per Hop Reservation (PHR) protocol. The signaling messages used by each protocol are given as well.

Per Domain Reservation – PDR Protocol

The PDR protocol manages the reservation of the resources in the entire Diffserv domain and is only implemented in the edge nodes of the domain. This protocol handles the interoperation with external resource reservation protocols and PHR protocol. The PDR protocol thus can be seen as a link between the external resource reservation scheme and the PHR.

The linkage is done at the edge nodes by associating the external reservation request flow identifier (ID) with the internal PHR resource reservation request. This flow ID, depending on the external reservation request, can be of different formats. For example, a flow specification ID can be a combination of source IP address, destination IP address and the DSCP field.

A PDR protocol has a set of functions associated, regardless of whether PDR protocol is an existing protocol or a newly-defined protocol. But, depending on the type of network where RMD is applied, it may have also a specific set of functions.

A PDR protocol implements all or a subset of the following functions:

- Mapping of external QoS requests to a Diffserv Code Point (DSCP);
- Admission control and/or resource reservation within a domain;
- Maintenance of flow identifier and reservation state per flow (or aggregated flows), e.g. by using soft state refresh;
- Notification of the ingress node IP address to the egress node;
- Notification that signaling messages (PHR and PDR) were lost in the communication path from the ingress to the egress nodes;
- Notification of resource availability in all the nodes located in the communication path from the ingress to the egress nodes;
- Severe congestion handling. Due to a route change or a link failure a severe congestion situation may occur. The egress node is notified by PHR when such a severe congestion situation occurs. Using PDR, the egress node notifies the ingress node about this severe congestion situation. The ingress node solves this situation by using a predefined policy, e.g., refuses new incoming flows and terminates a portion of the affected flows.

These functions are described in detail in [13].

Per Hop Reservation – PHR Protocol

The PHR protocol extends the PHB in Diffserv with resource reservation, enabling reservation of resources per traffic class in each node within a Diffserv domain. This protocol is not able to differentiate between individual traffic flows, as for example RSVP [10], as there is no per-flow information stored and no per-flow packet scheduling. Therefore, it scales very well.

The RMD framework defines two different PHR groups:

- **The Reservation-based PHR** group enables dynamic resource reservation per PHB in each node in the communication path. All the nodes maintain one state per PHB and no per-flow information. The reservation is done in terms of resource units, which may be based on a single parameter, such as bandwidth, or on more sophisticated parameters.
- **The Measurement-based Admission Control (MBAC) PHR** group is defined such that the availability of resources is checked by means of measurements before any reservation requests are admitted, without maintaining any reservation state in the nodes in the communication path. These measurements are done on the average real traffic (user) data load.

Only one PHR protocol has been specified thus far, the RMD On-demand (RODA) PHR [14] protocol. RODA is a reservation-based unicast edge-to-edge protocol designed for a single DiffServ domain, aiming at extreme simplicity and low cost of implementation along with good scaling properties. The RODA PHR protocol is implemented on hop-by-hop basis on all the nodes in a single Diffserv domain. The resource reservation request signaled by RODA is based on a resource unit. A resource unit is a bandwidth parameter that must be reserved by all nodes in the communication path between ingress and egress. The edge nodes, i.e., ingress and egress, of the Diffserv domain use certain message types to request and maintain the reserved resources for the flows going through the Diffserv domain. Each flow can occupy a certain number of resource units assigned to a particular Diffserv class.

The RODA PHR protocol implements all or a subset of the following functions:

- Admission control and/or resource reservation within a node;
- Management of one reservation state per PHB by using a combination of the reservation soft state and explicit release principles;
- Stores a pre-configured threshold value on maximum allowable resource units per PHB;
- Adaptation to load sharing. Load sharing allows interior nodes to take advantage of multiple routes to the same destination by sending via some or all of these available routes. The RODA PHR protocol has to adapt to load sharing when it is used;
- Severe congestion notification. This situation occurs as a result of route changes or a link failure. The PHR has to notify the edges when this occurs;
- Transport of transparent PDR messages. The PHR protocol may encapsulate and transport PDR messages from an ingress node to an egress node.

The sets of functions performed by the RODA protocol are described in [13], [14].

RMD Signaling Messages

The RMD signaling messages are categorized into RODA PHR and PDR protocol messages. These signaling messages and their description are depicted in Table 1.

The PDR signaling messages such as the "PDR_Reservation_Request", "PDR_Refresh_Request" or "PDR_Release_Request" messages may be encapsulated into a RODA PHR message or sent as separate messages. The PDR messages encapsulated into RODA PHR messages will contain the information that is required by the egress node to associate this RODA PHR signaling message with, for example, the PDR flow ID and/or the IP address of the ingress node.

Table 1. RODA PHR and PDR signaling messages

Protocol	Signaling Message	Signaling Message Description
RODA PHR	"PHR_Resource_Request"	Initiate the PHB reservation state on all nodes located on the communication path between the ingress and egress nodes according to the external reservation request.
	"PHR_Refresh_Update"	Refreshes the PHB reservation soft state on all nodes located on the communication path between the ingress and egress nodes according to the resource reservation request that was successfully processed by the PHR functionality during a previous refresh period. If this reservation state does not receive a "PHR_Refresh_Update" message within a refresh period, reserved resources associated to this PHR message will be released automatically.
	"PHR_Release_Request"	Explicitly releases the reserved resources for a particular flow from a PHB reservation state. Any node that receives this message will release the requested resources associated with it, by subtracting the amount of PHR requested resources from the total reserved amount of resources stored in the PHB reservation state
PDR	"PDR_Reservation_Request"	Initiates or updates the PDR state in the egress. It is generated by ingress node.
	"PDR_Refresh_Request"	Refreshes the PDR states located in the egress. It is generated by the ingress node.
	"PDR_Release_Request"	Explicitly release the PDR state. It is generated by the ingress node. Applied only when the PDR state does not use a reservation soft state principle.
	"PDR_Reservation_Report"	Reports that a "PHR_Resource_Request"/"PDR_Reservation_Request" has been received and that the request has been admitted or rejected. ". It is sent by the egress to the ingress node.
	"PDR_Refresh_Report"	Reports that a "PHR_Refresh_Update"/"PDR_Refresh_Request" message has been received and has been processed. It is sent by the egress to the ingress node.
	"PDR_Congestion_Report"	Used for severe congestion notification and is sent by egress to ingress. These PDR report messages are only used when either the "greedy marking" or "proportional marking" severe congestion notification procedures are used.
	"PDR_Request_Info"	Contains the information that is required by the egress node to associate the PHR signaling message that encapsulated this PDR message to for example the PDR flow ID and/or the IP address of the ingress node. It is generated by the ingress node.

RMD Functional Operation

The functional operation of the RMD framework given here is based on the interoperation between the RODA PHR and PDR protocol functions, assuming that PDR is a newly defined protocol. Two illustrative functional operation examples are presented:

- Normal Operation that describes the scenario for successful reservation between edges of a Diffserv domain
- Fault handling that describes the severe congestion handling between edges of a Diffserv domain

Normal Operation

When an external "QoS Request" arrives at the ingress node (see Fig. 1), the PDR protocol, after classifying it into the appropriate PHB, will calculate the requested resource unit and create the PDR state. The PDR state will be associated with a flow specification ID. If the request is satisfied locally, then the ingress node will generate the "PHR_Resource_Request" and the "PDR_Reservation_Request" signaling message, which will be encapsulated in the "PHR_Resource_Request" signaling message. This PDR signaling message may contain information such as the IP address of the ingress node and the per-flow specification ID. This message will be decapsulated and processed by the egress node only.

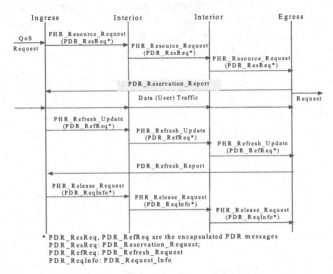

Fig. 1. RMD functional operation for a successful reservation

The intermediate interior nodes receiving the "PHR_Resource_Request" must identify the Diffserv class PHB (the DSCP type of the PHR signaling message) and, if possible, reserve the requested resources. The node reserves the requested resources by adding the requested amount to the total amount of reserved resources for that Diffserv class PHB.

The egress node, after processing the "PHR_Resource_Request" message, decapsulates the PDR signaling message and creates/identifies the flow specification ID and the state associated with it. In order to report the successful reservation to the ingress node, the egress node will send the "PDR_Reservation_Report" back to the ingress node. After receiving this report message, the ingress node will inform the external source of the successful reservation, which will in turn send traffic (user data).

If the reserved resources need to be refreshed (updated), the ingress node will generate a "PHR_Refresh_Update" to refresh the RODA PHR state and

"PDR_Refresh_Request" message to refresh the PDR soft state in the egress node. The PDR refresh message will be encapsulated in the "PHR_Refresh_Update".

Apart from the soft state principle, the reserved resources in any node can also be released explicitly by means of explicit release signaling messages. In this case, the ingress node will create a "PHR_Release_Request" message, and it will include the amount of the PHR requested resources to be released. This message will also encapsulate a PDR information message. The amount of the resources to be released will be subtracted from the total amount of reserved resources in the RODA PHR state.

If there were no resources available in one of the interior nodes (see Fig.2), the "PHR_Resource_Request" will be "M" marked and, as a result, the reservation request will be rejected. The ingress node will be notified of the lack of the resources by means of the "M" marked "PDR_Reservation_Report" message. Furthermore, in addition to marking the "PHR_Resource_Request" message, the interior node will also include the number of the interior nodes that successfully processed this message. This number can be derived from the TTL value in the IP header of the packet received. Using this information the ingress node will generate a "PHR_Release_Request" that will release the resources in the interior nodes that have reserved PHR resources for the rejected resource reservation request. In this way, the degradation of link utilization will be minimized.

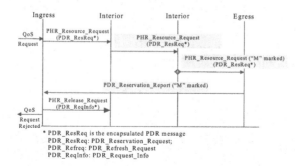

Fig. 2. RMD functional operation for a failed reservation

Fault Handling

Fault handling functional operation refers to handling undesired events in the network, such as a route changes, link failures, etc. This can lead to the loss of PHR signaling messages or to severe congestion.

Severe Congestion Handling

Routing algorithms in networks will adapt to severe congestion by changing the routing decisions to reflect changes in the topology and traffic volume. As a result the re-routed traffic will follow a new path, which may result in overloaded nodes as they need to support more traffic than their capacity allows. This is a severe congestion occurrence in the communication path, and interior nodes need to send notifications to the ingress nodes by means of PHR and PDR signaling messages. The ingress node has to resolve this situation by using a predefined policy. The interior node first detects the severe congestion occurrence, after which it will first notify the egress node and subsequently the ingress node. One can think of various detection and notification methods for the interior nodes, such as marking of all the data packets passing through a severe congested node or marking the PHR signaling messages only. In this paper only one method is considered. This is the "proportional marking" method, where the number of the remarked packets is proportional to the detected overload. The severely congested interior node will remark the user data packets with a domain specific DSCP (see [4]), proportionally to the detected overload. Once the marked packets arrive at the egress node, the egress node will generate a "PDR_Congestion_Report" message that will be sent to the ingress node. This message will contain the over-allocation volume of the flow in question, e.g., a blocking probability.

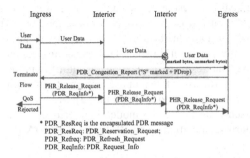

Fig. 3. RMD functional operation in case of severe congestion

For each flow ID, the egress node will count the number of marked bytes and the number of unmarked bytes, and it will calculate the blocking probability using the formula (1):

$$P_{drop} = \frac{B_m}{B_m + B_u} \quad \textbf{(1)}$$

where B_m = number of marked bytes and B_u = number of unmarked bytes.

The ingress node, based on this blocking probability, might terminate the flow. That is, for a higher blocking probability there is a higher chance that the flow will be terminated. If a flow needs to be terminated, the ingress node will generate a "PHR_Release_Request" message for this flow.

RMD Performance Evaluation

The performance behavior of the RMD framework has been studied through performance evaluation. For this evaluation, simulation and performance measurement techniques were used. These experiments focus on the two main RMD operation scenarios, the normal operation and the fault handling, described in Sections 0, 0 respectively.

Normal Operation

The normal operation of RMD (see Section 0) was simulated using a network simulator (ns) [20] environment. The goal was to study the RODA PHR protocol efficiency on bandwidth utilization. The experiments were performed on the RODA PHR both with and without using explicit release signaling messages ("PHR_Resource_Release" message - see Section 0). The RMD normal operation with the explicit release of resources is depicted in Fig.1, while the RMD normal operation without explicit release of resources is described in [13][14]. In the case of the RODA PHR without explicit release, the resources will be released based on the soft state principle after a refresh timeout. This has an impact on resource utilization. In order to improve the resource utilization, a sliding window is used. A sliding window enables faster release of the unused reservations compared to the refresh timeout. This algorithm is explained in detail in [13][14].

   This section presents the simulation model, the performance experiments and their results for RMD normal operation, for both types of RODA PHR, with and without explicit release.

Simulation Model

The network topology used for performance evaluation of RMD normal operation is shown in Fig. 4.. The link capacities, i.e. bandwidth, between the routers are set to C, 2xC and 3xC, where C=620 units. A single resource unit is set to 2000 bytes/second rate allocation. This particular value was chosen as it represent the rate required by an encoded voice communication, e.g., GSM coding. There are 3 sources generating the same type of traffic to a single destination. Note that in this model there was only signaling traffic generated, thus there was no actual data load. As all three sources generate the same type of traffic, only one traffic class is used. That is, the resource requests are related to a single DSCP.

   The resource requests ("PHR_Resource_Request"s – see Fig. 4) are generated according to Poisson process, with intensity of 1.356 requests/second. The requests holding times are distributed exponentially with mean set to $\frac{1}{\mu} = 90$ seconds. The resource unit request is distributed evenly between 1 and 20 units. With these traffic and bandwidth parameters the theoretical request blocking probability is 50%.

   The length of the soft state refresh period is set to 30sec. Using the sliding window (see [15], [17]), in case of RODA PHR without explicit release, the length of the refresh timeout can be decreased on average, from 1.5 times to (1 + 1/N) times of the refresh period, where N denotes the number of cells in the sliding window. In the simulation model, the number of cells in a refresh window is set to 30, i.e. N = 30.

The expected output of the performance experiments with this simulation model were the results on link reservation and link utilization. The link reservation represents the number of the reserved resources per traffic class, i.e. Diffserv class on a link. The link utilization represents the amount of resource that can be actually used by traffic sources. For example, the link 2-3 (see Fig. 4) utilization is the sum of the resources that can be used by source 0, 1 and 2.

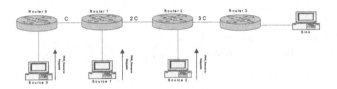

Fig. 4. Network simulation layout

Numerical Evaluation

Fig.5 and Fig. 6 present the reservation and utilization between routers 0 and 1 using the RODA PHR with and without explicit resource release. The [%] in the Fig.5 and Fig. 6 is the percentage of maximum reserved units available that are reserved/utilized. The maximum number of units is 620 on the first link, 2*620 on the second link, and 3*620 on the third link.

It can be seen that link reservation in both cases is slightly below 100 percent of link capacity, 94% and 91% respectively. However, the link utilization is improved by an average of 18% (from 62% up to 80%) when the explicit release messages are used when compared to the situation where there is no explicit release of resources. This is because, in the latter case, the resources are released based on a sliding window in proportion to the refresh period, while explicit release of resources is done in the order of roundtrip times of signaling messages. This round trip time is of course shorter than the length of the refresh period.

Fig.7 and Fig.8 show the results obtained for the link between router 2 and 3. Here again, link reservation is around 100 percent (97% and 98%) and link utilization is improved from 83% to 97% due to explicit resource release.

The average link utilization is higher on link between routers 2 and 3 than on the link between routers 0 and 1, because there are no downstream routers to reject a request already accepted by router 3. This is, however, not the case for the link between routers 0 and 1 as, for example, router 2 may reject resource request admitted by router 1. For the RODA PHR without explicit release of resources, it gets worse as the resources are kept reserved on average for the length of an additional 1.5 refresh periods. This means that the resource requests admitted by router 0 but rejected by routers 1 or 2 install reservation state for the length of 1.5 refresh periods, on average, even though the resource request is ultimately rejected.

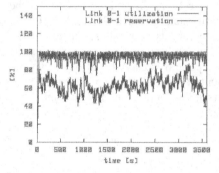

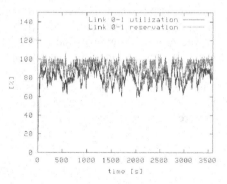

Fig. 5. Link utilization and reservation on link Router0-Router1 (RODA without release)

Fig. 6. Link utilization and reservation on link Router0-Router1 (RODA with release)

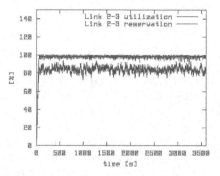

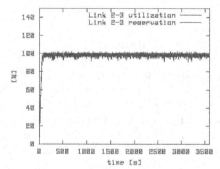

Fig. 7. Link utilization and reservation on link Router2-Router3 (RODA without release)

Fig. 8. Link utilization and reservation on link Router2-Router3 (RODA with release)

Performance Measurements

This section presents the performance measurements that were used to estimate the mean processing delays of the RODA PHR signaling messages. In order to compare the mean processing delays of the RODA PHR signaling messages with the mean processing delays of standardized IETF resource reservation protocols, performance measurements on the mean processing delays of the RSVP signaling messages are also done. For a 'fair' comparison between the two protocols, all experiments were performed using the same hardware.

The measurement topology consisted of two edge nodes interconnected by one interior node. The edge and interior nodes were PCs with 400Mhz Intel Pentium-II processors and 64 MBytes of RAM (Random Access Memory) running the Linux operating system. Furthermore, all nodes were running a preliminary prototype implementation of the RODA PHR protocol and a public implementation of the RSVP protocol [22]. Note that this preliminary prototype implementation of the

RODA PHR protocol does not support the processing of the PHR_Resource_Release message.

The ingress node sends either RODA PHR or RSVP signaling messages to the egress node in order to reserve and/or refresh resources for different flows in the interior node. During the performance measurements only one flow (session) was running on the interior node.

Fig. 9 depicts the measurement points located within the interior node (see also [21].). The measured processing delays are the processing delays in the (unloaded) interior node of the RODA PHR and RSVP protocol messages, excluding the delays necessary for the forwarding of these protocol messages through the IP protocol stack available in the node. The traffic is forwarded through the IP packet forwarder and Network Interface Card (see Fig. 9). The measured mean processing delay of the traffic forwarder is 12 µsec.

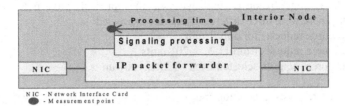

Fig. 9. Interior Node Measurement Points

The mean processing delays and their 95% confidence intervals of the RODA PHR and RSVP RESV signaling messages are listed in Table 2.

RSVP, which is a receiver-initiated reservation protocol, uses the RSVP PATH message to store backward routing information. This is used by the associated RSVP RESV message traveling back to the sender. The RMD protocol, which is sender-initiated, has no corresponding message. Therefore, the mean processing delay of the RSVP PATH message is not shown Table 2. However, its mean processing delay has been measured and is 273 [µsec].

Table 2. Mean processing delays and their 95% confidence intervals

Protocol Messages	RODA PHR processing delay (µsec)	RSVP processing delay (µsec)
Reservation message	PHR_Resource_Request 0.58 [0.0 ; 2.0]	RESV 790 [762.8 ; 793.2]
Refresh message	PHR_Refresh_Update 0.5 [0.0 ; 2.0]	RESV 67.02 [64.03;69.23]

From Table 2, it can be seen that the mean processing delays of the RODA PHR reservation messages are approximately 1338 times smaller than the mean processing delays of the RSVP reservation messages. The reason is that, unlike RSVP, the

RODA PHR does not maintain per-flow traffic classifiers, and it does not require per-flow maintenance and lookup in a reservation table.

Fault Handling

This section presents the performance behavior of the severe congestion handling described in Section 0. For this purpose, the RMD was simulated using the network simulator (ns) [20] environment. This section describes the simulation model used, the experiments performed and the results.

Simulation Model
Based on the operational description of the RMD protocol, resources are requested in bandwidth units, which are also used for the description of the traffic models. As in Section 0, a single resource unit was set to represent 2000 bytes/second rate allocation. There were three different scenarios examined:
  i)   resource requests for only 1 unit
  ii) resource requests from {1,2,...,20} units
  iii) resource requests selected from {1,2,...,100} units.

The resource requests generation model is a Poisson process. The average resource request call holding time was set to $\frac{1}{\mu} = 90$ seconds.

The resource requests were generated in such a way that the requested bandwidth for each reservation unit class was balanced, i.e.,

$$\frac{\lambda_i}{\mu} BW_i = \frac{\lambda_j}{\mu} BW_j \quad (2)$$

where $BW_i = i\, U$ [unit] is the bandwidth request, U = 2000 bytes/sec and $\frac{1}{\lambda_i} = 0.9$ sec

as per default. Hence, higher bandwidth requests arrived less frequently than smaller ones. Packet sizes (L) of the connections were determined according to their reservations:

$$L_i = L_1 * BW_i \quad (3)$$

where, L_1 = 40 bytes, is the packet size for a single unit connection $BW_1 = 1*U$ (assuming the packet inter-arrival time remains constant, which is $L_i/BW_i = $ const)
For the sake of simplicity, packet inter-arrival times were kept constant (constant bit rate - CBR), hence every flow sent one single packet in each 20 msec interval.
    For the evaluation of the methods, a simple delta network topology (see Fig. 10) was used.

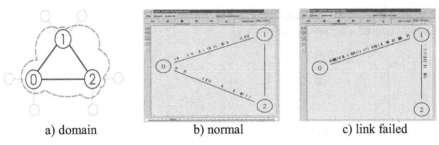

| a) domain | b) normal | c) link failed |

Fig. 10. Network Layouts

For the sake of simplicity, node 0, 1 and 2 represent the Diffserv domain using the RMD protocols. These nodes act as interior nodes as well as network edge nodes whenever necessary. In order to have effective multiplexing of flows, the bandwidth of the links (*C*) (see (4)) was set to be able to accommodate at least 100 flows of the highest bandwidth demands, i.e., to 100, 2000 and 100000 units.

$$C = 100 * BW_{maxi} \quad (4)$$

As discussed in Section 0, the severe congestion detection can be based on packet drop ratio measurements. Hence, it was important to find the proper dimensioning for the network buffers. As this traffic model was based on CBR traffic with 20 msec packet inter-arrival times, the queue lengths were set such that no packet loss can occur during normal operation. The buffer sizes (*B*) were determined using the following formula:

$$B = C * \frac{L_i}{BW_i} * C_{threshold} \quad (5)$$

where, $C_{threshold}$ is the amount of maximum amount of resources available for a single traffic class.

The dimensioning was done for a target load level of 80% link capacity, hence buffer size is set to $B=C*0.02*0.8$ [bytes], assuming the 20 msec packet inter-arrival time.

In order to model a failure event in the network in **Fig. 10**, after the system achieves stability, the link between nodes 0 and 1 goes down at 350 sec of simulation time. Afterwards, the dynamic routing protocol (OSPF) updates its routing table at 352.0 sec and all flows previously taking the 0-2 path will be re-routed to the 0-1-2 alternate path. Due to this, node 0 will suffer a severe overload resulting from the re-route event.

Numerical Evaluation

The detailed simulation results obtained for severe congestion handling described in Section 0 are presented below.

After detecting the severe congestion situation, the severe congestion handling algorithm immediately drops some of the flows in proportion to the detected overload. Thereby, the load of the link between nodes 0 and 1 almost instantly returns to the target load level of 80 percent. (see Fig. 11). From Fig. 12 one can see the short-term transients caused after severe congestion.

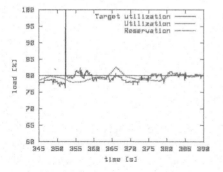 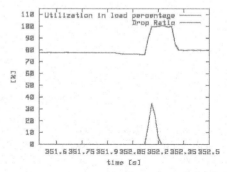

Fig. 11. Utilization and reservation after severe congestion

Fig. 12. Short term transients in packet drops and utilization during severe congestion

It is a natural side effect that during congestion the highly saturated queues will induce shifted round trip times for the connections (see Fig. 13). It can be seen that it takes about 2 seconds for the dynamic routing protocol[1] to update its link state and re-direct the affected connection. It can also be seen that the retained round trip time doubles for connections now traversing along 2 hops (every link's delay is 5 msec). Note that these times are extremely low (under 26 msec). That is, they correspond to one or two voice frames, which is quite acceptable.

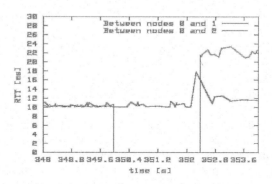

Fig. 13. The transient in Round Trip Times

Conclusions

In this paper an overview of the RMD functionality and its performance behavior is given. RMD is a simple, fast, low-cost and scalable dynamic resource reservation scheme. As such, it enhances the already scalable Diffserv architecture with dynamic

---

[1] We added the Open Shortest Path First (OSPF) routing extension to ns.

resource reservation and admission control. In order to observe the overall RMD functionality behavior and to prove the RMD concepts, extensive performance evaluation experiments have been performed.

When operating under normal conditions, the results show that network utilization is very close to maximum achievable even though the protocol does not use per-flow state maintenance in core routers. The performance measurement experiments show that the mean processing delays of RODA PHR reservation messages are more than 1338 times smaller than the mean processing delays of RSVP reservation messages. The performance evaluation of the severe congestion procedure has shown that the RMD reaction and recovery time on such events is negligible. Furthermore, using the RMD concepts, the utilization of the links is only slightly affected by severe congestion occurrence.

The performance behavior of RMD will be further investigated by experimenting with network topologies consisting of a larger number of nodes. Moreover, comparison with other types of RSVP implementation distributions such as e.g. KOM RSVP engine [23] could be accomplished.

Also there are still open issues with RMD that need to be studied, such as for example, extending the RMD applicability in a multi-domain, extending RMD with policy control and the Measurement-based PHR that still needs to be specified and evaluated.

Acknowledgements

Special thanks to our colleagues Simon Oosthoek, Pontus Wallentin, Martin Jacobsson, Marcel de Kogel, Geert Heijenk and Gábor Fodor for their useful inputs and comments.

References

1. Braden, R., Clark, D., Shenker, S., "Integrated Services in the Internet Architecture: an Overview", IETF RFC-1633, Jun. 1994
2. Wroclawski, J., " The use of RSVP with IETF integrated Services", IETF RFC 2210, 1997. "
3. Blake, S., Black, D., Carlson, M., Davies, E., Wang, Zh., Weiss, W., "An Architecture for Differentiated Services", IETF RFC 2475, 1998.
4. Nichols, K., Blake, S., Baker, F. and D. Black, "Definition of the Differentiated Services Field (DS Field) in the IPv4 and IPv6 Headers", RFC 2474, December 1998.
5. Preliminary WWW site of the IETF NSIS WG located at: http://www1.ietf.org/mailman/listinfo/nsis
6. Braden, R., Zhang, L., Berson, S., Herzog, A., Jamin, S., "Resource ReSerVation Protocol (RSVP)-- Version 1 Functional Specification", IETF RFC 2205, 1997.
7. Baker, F., Iturralde, C. Le Faucher, F., Davie, B., "Aggregation of RSVP for IPv4 and IPv6 Reservations", IETF RFC 3175, 2001.
8. Bergkvist, J., Cselényi, I., Ahlard, D., "Boomerang – A simple Resource Reservation Framework for IP", Internet Draft, draft-bergkvist-boomerang-framework-01.txt, Work in Progress, November 2000.

9. Pan, P., Schulzrinne, H., "YESSIR: A Simple Reservation Mechanism for the Internet", Proceedings NOSSDAV'98, 1998.
10. Chow, H., Leon-Garcia, A., "A Feedback Control Extension to Differentiated Services", Internet Draft, draft-chow-diffserv-fbctrl-01.txt, Work in Progress, March 1999.
11. Stoica, I., Zhang, H., Shenker, S., Yavatkar, R., Stephens, D., Malis, A., Bernet, Y., Wang, Z., Baker, F., Wroclawski, J., Song, C., Wilder, R., "Per Hop Behaviours Based on Dynamic Packet States", Internet Draft draft-stoica-diffserv-dps-00.txt, Work in Progress, February 1999.
12. White, P.P., Crowcroft, J., "A Dynamic Sender-Initiated Reservation Protocol for the Internet", Proceedings, HPN'98, 1998.
13. Westberg, L., Jacobsson, M., Karagiannis, G., Oosthoek, S., Partain, D., Rexhepi, V., Szabo, R., Wallentin, P., "Resource Management in Diffserv Framework", Internet Draft, Work in Progress, 2001.
14. Westberg, L., Karagiannis, G., Partain, D., Oosthoek, S., Jacobsson, M., Rexhepi, V., "Resource Management in Diffserv On DemAnd (RODA) PHR", Internet Draft, Work in progress.
15. Jacobsson, M., "Resource Management in Differentiated Services – A Prototype Implementation", M.Sc. Thesis, Computer Science/TSS, University of Twente, June 2001.
16. Heijenk, G., Karagiannis, G., Rexhepi, V., Westberg, L., "DiffServ Resource Management in IP-based Radio Access Networks", Wireless Personal Multimedia Communications (WPMC'01), Aalborg, Denmark, September 2001.
17. Marquetant, A., Pop, O., Szabo, R., Dinnyes, G., Turanyi, Z., "Novel enhancements to load control a soft state, lightweight admission control protocol", QofIS'2000 - 2nd International Workshop on Quality of future Internet Services, September 2001.
18. Csaszar, A., Takacs, A., Szabo, R., Rexhepi, V., Karagiannis, G., "Severe Congestion Handling with Resource Management in Diffserv On Demand", submitted to Networking 2002, May 19-24 2002, Pisa - Italy.
19. Partain, D., Karagiannis, G., Westberg, L., "Resource Reservation Issues in Cellular Access Networks", Internet Draft, Work in progress.
20. The Network Simulator - ns-2, http://www.isi.edu/nsnam/ns/
21. Feher, G., Korn, A., "Performance Profiling of Resource Reservation Protocols", IFIP 2001, 27th - 29th June, 2001,Budapest- Hungary.
22. ISI public source code for the RSVP protocol, located at (www.isi.edu/div7/rsvp/)
23. KOM RSVP engine located at (http://www.kom.e-technik.tu-darmstadt.de/rsvp/)

Performance Evaluation of the Extensions
for Control Message Retransmissions in RSVP

Michael Menth* and Rüdiger Martin

Department of Distributed Systems, Institute of Computer Science, University of Würzburg,
Am Hubland, 97074 Würzburg, Germany,
{menth,martin}@informatik.uni-wuerzburg.de

Abstract. Quality of Service (QoS) for real-time transmission can be achieved by resource reservation in the routers along the path. In recent years, several protocols and extensions to them have been designed for signaling resource reservation in IP networks. This work reviews various protocols that exhibit different signaling concepts. Then, we study the impact of control message retransmissions (CMR) and the control option on the reservation establishment delay (RED) and the reservation teardown delay (RTD). Numerical results quantify the resulting performance gain in different networking scenarios.

1 Introduction

Future communication networks will guarantee seamless quality of service (QoS) data transportation from the sender to the receiver. The network must provide sufficient resources to forward the data in an adequate way to meet the loss and delay requirements of the traffic. To achieve that goal, massive overprovisioning can be applied as well as intelligent traffic engineering techniques. One of them is admission control (AC): When the network's capacity does not suffice to transport all offered traffic, AC shelters the network from overload by admitting only a limited number of reservation requests. Thus, QoS for the flows in place is maintained at the expense of blocked flows. In order to perform AC in a network entity, the amount of requested resource of each flow must be known beforehand and is usually delivered by a resource reservation protocol.

In recent years, several resource reservation protocols have been designed for IP networks. In this paper we would like to give an overview over the most prominent protocols: RSVP, RSVP refresh overhead reduction extensions, aggregation of RSVP reservations, Boomerang, YESSIR, BGRP, and stateless reservation protocols. These protocols distinguish in syntax and semantics and reveal also different information passing concepts.

In the past, protocol performance has been studied using software implementations [1,2,3,4]. In [5] the reliability of RSVP was studied. In our investigation we focus on the control message retransmission (CMR) option which is suggested in the refresh overhead reduction extensions of RSVP [6]. We evaluate its impact on the reservation

---

* This work was partially funded by the Bundesministerium für Bildung und Forschung of the Federal Republic of Germany (Förderkennzeichen 01AK045). The authors alone are responsible for the content of the paper.

G. Carle and M. Zitterbart (Eds.): PfHSN 2002, LNCS 2334, pp. 35–49, 2002.

establishment delay (RED) and the reservation teardown delay (RTD). We perform these studies in various networking scenarios and take signaling paradigms from standard RSVP and other reservation protocols into account.

This paper is structured as follows. In Section 2 we present the above mentioned protocols. Then, we study the response time of resource reservation protocols with special respect to the control message retransmission (CMR) feature in the RSVP extensions. Numerical results illustrate their behavior and quantify the performance gain. In Section 4 we summarize this work.

2   An Overview of Resource Reservation Protocols

Real-time applications like voice over IP (VoIP) [7] or video conference require signaling protocols for the application layer as well as for the network and the transport layer.

Applications need to identify and locate their communication peers and to negotiate session parameters. Codecs have to be agreed on and translators can be involved in case of incompatible end systems. These and other tasks are performed by standards like H.323, the Session Initiation Protocol (SIP), and the Media Gateway Control Protocol (MGCP). The Real-time Transport Protocol (RTP) provides means for data synchronization to avoid distorted time lines for presentation. On the transport layer we have the User Datagram Protocol (UDP) and the Transmission Control Protocol (TCP). They ensure that IP packets are associated with the correct ports in the end systems, and TCP provides means to detect and repair packet loss. The network layer forwards the data to their destination. It consists of the Internet Protocol (IP) and the routing protocols.

The link layer offers QoS mechanisms as it has the control over the overall capacity of a link. These resources do not only relate to mere bandwidth but also to CPU processing time, buffer space, and others. Resource reservation can be performed per hop on all intermediate links between sender and receiver. The control messages are transported using regular IP packets (network layer) but their information relates to the link layer.

This work focuses on signaling for resource reservation. In this section we present various existing protocols and concentrate on their information forwarding paradigm.

2.1   Resource Reservation Protocol (RSVP)

RSVP has been conceived by the IETF for the signaling of reservation requests within an Integrated Services network [8,9]. Both unicast and multicast applications are supported and different reservation styles (e. g. shared reservations) are possible.

Connection Establishment. To initiate a reservation with RSVP, the sending node issues a so-called PATH message that establishes a PATH state (PS) in the intermediate hops on the way to the desired destination machine. The state of a flow is the information related to it in a router. The destination router responds with a RESV message that visits the intermediate routers in the reverse direction using the previous hop information of the PS (cf. Figure 1). On that way, the RESV states (RS) are established. This ensures that resources are reserved in each router in downstream direction. The first pass from the sender to the receiver (PATH msg.) collects advertising information that is delivered to

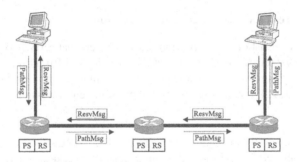

Fig. 1. Control message flow in RSVP.

the receiver to enable the receiver to make appropriate reservation requests. The actual reservation is made on the way back to the sender (RESV msg.). Hence, RSVP uses a two-pass signaling approach, also known as one-pass with advertising (OPWA). Explicit PATHERR and RESVERR messages indicate errors, and TEARDOWN messages tear down the connection and remove the states in the routers.

When a RESV message is received by a router, the required actions are taken to set up the reservations for the respective data flow (cf. Figure 2). The admission decision is based on the flow and reservation specifiers. If the request succeeds, the classifier and the scheduler are configured to forward the data flow messages.

Soft States. RSVP control messages are usually sent directly in IP or UDP datagrams. This communication is unreliable in both cases. The control messages do not belong to a user data flow and are not protected by the reservation. In addition, the end systems may go down without notifying the network. Thus, a mechanism is required to remove the unused states. RSVP uses a soft state approach to cope with that: The states time out and disappear after a cleanup time L unless they are refreshed by another PATH or RESV message. To keep the connection alive, every participating node sends periodically PATH and RESV messages to its neighboring hops with refresh period R. RSVP [8] suggests to set L to $3 \cdot R$. If the source stops without tearing down the connection or in case of routing changes, the PATH and RESV states will eventually time out in all unused routers.

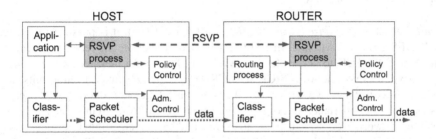

Fig. 2. The establishment of a reservation using RSVP.

2.2 RSVP Refresh Overhead Reduction Extensions

RSVP messages consist of different standardized objects that carry flow related information. The RSVP refresh overhead reduction extensions [6,10] introduce several new objects to RSVP that reduce the overhead due to the transmission of refresh messages. Apart from that, RSVP is enhanced by control message retransmission (CMR) capabilities.

BUNDLE Messages. RSVP nodes send one PATH and RESV message per refresh interval and connection. Since several connections are carried on the same link, their refresh messages can be handled within one single BUNDLE message where just the different message bodies are assembled. This yields just a reduction of the mere control packet frequency but not a reduction of the control message frequency. The bandwidth consumed by control messages is negligible and BUNDLE messages hardly reduce it. There are also as many operations required by the router as in normal RSVP.

Complexity Reduction by MESSAGE_IDs. PATH and RESV messages are sent periodically per RSVP connection and their content rarely changes. The receiver of such a message identifies the corresponding flow using the traffic descriptor and refreshes or updates its state. The introduction of a MESSAGE_ID object alleviates this task. If a control message is expected to change the state in the next router, it is equipped with a unique MESSAGE_ID number. Consecutive control messages that just refresh this state are equipped with the same MESSAGE_ID. The desired state is then identified by a hash value using the MESSAGE_ID and can be refreshed without processing the whole control message. MESSAGE_IDs speed up the lookup time for a flow. The impact on the flow processing time can be studied by comparing the performance of RSVP implementations.

Summary Refresh Extensions. The MESSAGE_IDs do not require to be sent with their related control message. A MESSAGE_ID LIST object may contain only MESSAGE_ID objects instead of the whole control message that would only be used in the failure case. This reduces also the required bandwidth for signaling, however, this has no impact on the network utilization since the fraction of signaling traffic is small anyway. To handle cases where the receiver encounters an inconsistent state view, the receiver may order new PATH or RESV messages by issuing a MESSAGE_ID_NACK that refers to the corrupted RSVP connection.

Control Message Retransmissions (CMR). Last but not least, it is possible to set an ACK_DESIRED flag within a MESSAGE_ID object to indicate the receiver to send a MESSAGE_ID_ACK to acknowledge the receipt of a control message. If the sender has not yet received the MESSAGE_ID_ACK object after R_f time, it retransmits the respective control message. This makes the communication more robust against packet loss which is an important issue, e. g., in wireless networks. To adjust this mechanism to potential network overload situations, an exponential backoff algorithm is introduced to avoid unnecessary control messages. On the one hand, CMR make RSVP implementations more complex (additional timers and control messages) but on the other hand, they make the communication on a lossy link more reliable which reduces the response time of the distributed RSVP processes. The impact depends on the networking scenario and will be investigated in the second part of that work.

2.3 Aggregation of RSVP Reservations

The above mentioned modifications to RSVP tend to reduce the protocol overhead per RSVP control message and allow better performing implementations of the RSVP state machine. However, they are not able to solve the fundamental scaling problem: The processing costs in a router grow linearly with the number of supported reservations which is feasible in the access network with only a few QoS flows but not in a core network. Therefore, [11] suggests an aggregator at a border router of a network that summarizes many individual RSVP reservations into one aggregated reservation. The aggregated reservations share the same path through the network and they are deaggregated at the egress point. This reduces the number of reservation states drastically within the network and relieves the core routers from processing individual reservation requests. The number of end-to-end tunnel reservations rises quadratically with the number of aggregators and deaggregators in a network. The same objective can be achieved by using the aggregation capabilities of Multiprotocol Label Switching (MPLS) [12,13,14].

2.4 Boomerang

The Boomerang protocol [15] aims at reducing some part of the overhead that is induced by RSVP. There are no PATH and RESV messages. The sender generates a Boomerang message that is forwarded hop by hop to the receiver. Along that path, the routers that understand Boomerang perform a reservation for a flow. As soon as the message arrives at the receiver, the reservation is already in place. The receiver does not even need to process the message, it just bounces the message back to the sender to notify the establishment. Optionally, the return channel of a bidirectional session may be reserved on the way back. Note that a different path may be taken for that purpose (cf. Figure 3).

If a reservation request fails or if a session terminates, the reservation states can be torn down with a reservation request of size zero. Only the sending node generates signaling messages, therefore, the Boomerang approach is simpler than RSVP since the major complexity and processing is located at the sender. The concept is also based on soft states and requires refreshes to keep the reservation alive. If the Boomerang message does not return to the initiating node within a certain time, it is considered to be lost, so

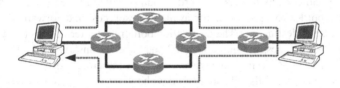

Fig. 3. A bidirectional reservation setup by Boomerang.

that the sender can take appropriate actions. The Boomerang protocol induces clearly less burden on the routers than conventional RSVP implementations [3].

2.5   YESSIR – YEt another Sender Session Internet Reservation

YESSIR [16] is a reservation protocol that is based on RTP [7,17]. RTP is usually a wrapper for application data and adds sequence numbers, time stamps and other identifiers. Each session is controlled by the Real-time Transport Control Protocol (RTCP). Senders and receivers send periodically sender and receiver reports (SR, RR). SRs contain throughput and other information about the last report interval and allow, e. g., the derivation of the current round-trip time in the network. RRs indicate packet loss and delay statistics among others. This is extremely useful for adaptive applications. YESSIR works like RSVP also in a unicast and a multicast environment and offers also different reservation styles.

YESSIR reservation messages are piggybacked at the end of RTCP SR or RR messages, possibly enhanced by additional YESSIR-specific data, carried in IP packets with router-alert option, i. e. they are intercepted by routers and processed by those supporting this option. As with Boomerang, reservations are triggered by the sender. If a router along the way is not able to provide the requested resources, the exact reasons for the reservation failure can be remarked. This helps the end systems to either drop the session or to lower the reservation request. The rate for the reservation can be given explicitly, it may be deduced from codec types in the RTP payload or it may also be inferred from the size of the payload and the corresponding time stamps. YESSIR also relies on the soft state approach. As with Boomerang, only the sender issues refreshes and the session can be torn down with an explicit RTCP BYE message. Unlike RSVP, the intermediate nodes are not able to issue error messages, failure situations have to be recognized by the receiver and reported via RRs to the sender.

2.6   Border Gateway Reservation Protocol (BGRP)

BGRP [18] has been conceived for inter-domain use and to work in cooperation with the Border Gateway Protocol (BGP) for routing. It is used for reservations between border routers only. BGRP addresses the scalability problem directly since it is designed to aggregate all inter-domain reservations with the same autonomous system (AS) gateway as destination into a single funnel reservation, no matter of their origin.

We explain briefly how signaling with BGRP works to set up a sink tree reservation (cf. Figure 4). A PROBE message is sent from a source border router to a destination border router and collects the visited border routers. Upon the reception of a PROBE message, the border routers check for available resources, and forward the PROBE packet towards the destination. The destination border router terminates this process. It converts the PROBE message into a GRAFT message and inserts an ID that identifies its sink tree. The GRAFT message travels back on the collected path: The required reservation states are established and marked with the ID, or they are updated if they are already existent. The PROBE and GRAFT messages contain only a relative reservation offset, therefore, the communication for GRAFT messages must be reliable (e. g. using

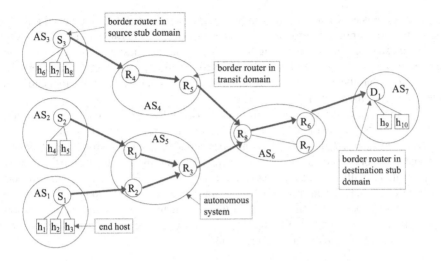

Fig. 4. Signaling in BGRP.

TCP). BGRP is a soft state protocol, therefore, neighboring routers exchange explicit REFRESH messages to keep the reservation alive.

Here, we have also receiver based reservations but in contrast to RSVP, the information is not stored per flow but per aggregate which is characterized by the same destination AS. Like with RSVP reservation aggregation, the advantage of that approach is the state scalability in the routers. In addition, there is only one sink tree reservation for every destination AS in each border router. Hence, the number of BGRP reservations scales linearly with the number of AS.

2.7 Measurement Based Admission Control

The above described reservation protocols suffer from a common disadvantage. They keep a state for every reservation, either for an individual or for an aggregate reservation. Their information is used to derive the already reserved capacity on the link and the AC decision is based on the remaining capacity.

As an alternative, measurement based AC (MBAC) gains this information by an estimation of the current network load by traffic measurements. Therefore, routers do not store any flow related information and MBAC has no scaling problems. However, MBAC might not be a good solution if strict guarantees are required. To support this architecture, the signaling protocols may be the same as in the conventional case, however, some special purpose protocols for MBAC have been also conceived [19,20,21]. MBAC is problematic if flows do not send continuously since this leads to wrong traffic estimations.

2.8 Stateless Reservation Protocols

A special MBAC approach are stateless reservation protocols. We only describe their basic architecture, for further details the reader is referred to [22,23,24].

End systems equip the packets that belong to reserved traffic with a special priority tag. Instead of keeping a record for different reservations, the intermediate routers analyze the packet streams on each outgoing link. They count the packets with the reservation tag within a given interval and infer the reserved rate R_{res}. For a correct measurement, this requires that the holder of a reservation sends also packets when the application is idle, otherwise, the reservation is reduced.

A sender can reserve resources from the network by sending request packets at a certain rate to the destination. The newly admitted rate R_{new} is also recorded over this interval. The sum of $R_{res} + R_{new}$ is an upper bound on the reserved rate on an output port and may not exceed the link capacity R_{link}. The intermediate hops drop the request packets intentionally if there is not enough capacity ($R_{res} + R_{new} > R_{link}$) to transport another flow. Otherwise, these packets are forwarded. A new reservation is only admitted if its request packets pass all AC tests in the intermediate routers and the destination signals the success back to the source. This is only the basic mechanism that does not reveal the manifold implementation problems.

3   Performance Evaluation of Control Message Retransmissions

Retransmissions for control messages (CMR) make the communication for RSVP control traffic more robust against packet loss. This is crucial in scenarios with high packet loss, e. g. in wireless networks. CMR lead to faster reactions of the RSVP processes in the distributed routers. The reaction time of the remote processes affects the reservation establishment delay (RED) and the reservation teardown delay (RTD). The establishment of a reservation is a prerequisite for the start of real-time applications, e. g. video streaming, and it is annoying for the user if he or she has to wait. Therefore, a short RED is important for good QoS perception by the user (post dial delay). Unused and blocked capacity is not profitable, therefore, resources should be released very quickly after session termination by the application layer. This requires a short RTD. In this section, we give a brief description of the options under study and illustrate their influence on RED and RTD.

3.1   Model Description

In our investigation we consider a general signaling protocol with different options. We borrow most of the nomenclature from RSVP but we do not limit our experiments to configurations in RSVP. We neglect the message processing times in the routers and focus on the effect of the mere transmission times and involved timeout values. The performance measures RED and RTD do not depend on the characteristics of the user traffic but on packet loss probability on a link and the path length. The analytical calculations are lengthy but straightforward, so we omit them in this presentation.

Control Message Retransmissions. We recall briefly the concept of CMR and point out the influencing parameters. The reservation process of a sender issues a control message and sets the retransmission timer. The receiver of this message is required to immediately return an acknowledgement. If the sender does not receive an acknowledgement before

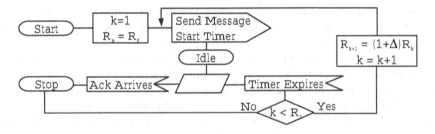

Fig. 5. A flowchart of the retransmission algorithm.

the timer expires, the control message is retransmitted. The retransmission timer value R_k depends on the kth retransmission interval: $R_k = (1 + \Delta)^{k-1} \cdot R_f$ for $1 \leq k < R_l$. R_k scales linearly with the rapid retransmission interval R_f. An exponential backoff is applied and Δ (we use $\Delta = 1$) governs the speed at which the sender increases the timer value. This avoids unnecessary retransmissions due to links with long transmission delays. The rapid retry limit R_l is an upper bound on the number of control message transmissions. A flowchart of the algorithm is depicted in Figure 5. This concept reduces only the response time of RSVP but it does not yield reliable communication. The parameter $R_l = 1$ corresponds to signaling without CMR. In case that no acknowledgement returns, the sender tries again after the refresh interval R. If a node has not received an update message after $3 \cdot R$, it faces a soft state timeout and sends a TEARDOWN control message to indicate the end of the session to its neighboring nodes.

Endpoint versus Common Control. In Boomerang or YESSIR, only the endpoints (sender, receiver) trigger control messages that are also processed in the other nodes. In RSVP, all participating routers control the connection, i. e. they do not only forward the control messages when they arrive, they also create refresh (PATH, RESV) messages when they do not receive them in time. We call the first approach "endpoint control" (EC) and the second one "common control" (CC). With CC and CMR, the ACKs are created and returned by neighboring hosts and not as under EC by the receiving peer over many intermediate hops. CC seems to make a reservation more robust against loss of control messages, especially in combination with CMR.

One-Pass versus Two-Pass. In RSVP, OPWA is used for establishing a connection which is in fact a two-pass approach: One pass is needed for setting up the PATH states in the routers and one pass back is required for setting up the reservations. The same holds for BGRP's PROBE and GRAFT messages. With Boomerang and YESSIR this is different. The reservation is done with the first pass from the sender to the receiver and a successful session setup may be notified to the source or not. Therefore, this is a true one-pass approach. With two-pass, the signaling takes on average twice as long as with one-pass. This is a relatively trivial result. To simplify the analysis, we concentrate only on the one-pass approach. As a consequence, the results for RED in our study must be doubled in case of RSVP OPWA.

Network Parameters. The effects of the retransmission timers depend certainly on the networking scenario. We make the following assumptions. We set the packet transmission delay per link to 10 milliseconds. The packet loss probability p_l on a single link influences the system as well as the number of links n in the reservation path. Therefore, we conduct studies varying theses parameters. If theses parameters are constant, we assume a path length of $n = 10$ hops and a link packet loss probability of $p_l = 10^{-2}$ which occurs in congested networks. For $n = 10$ hops, this yields an end-to-end packet loss probability of about 10%. Transatlantic routes can have more than 20 hops and the overall packet loss probability is sometimes more than 25%. We observed these values with the statistic tool for UDP traffic in realaudio or realvideo. Especially under these circumstances, reservations are crucial for real-time applications.

3.2   Reservation Establishment Delay

We are interested in the influence of the refresh interval R, the rapid retry limit R_l, and the control option on the RED.

First, we study the configuration EC without CMR as we find it in YESSIR. In Figure 6 the mean of the RED (E[RED]) is shown for a fixed path length ($n = 10$). For small link loss probabilities, the difference for various R is negligible. E[RED] rises with increasing loss rates and the difference between various values for R becomes visible. For high packet loss probabilities ($p_l = 10^{-2}$) the refresh interval R dominates RED almost linearly since R is several orders of magnitude larger than the packet transmission delay.

We set $p_l = 10^{-2}$ and observe the reservation for different path lengths. Figure 7 shows that the RED grows linearly with the path length. At first sight this seems to be a consequence of the summation of link transmission times but this is not the case because the size of a round trip time is in the order of hundred milliseconds. This phenomenon is rather due to the fact that a longer route exhibits a larger end-to-end loss probability. Because of the high loss probabilities, the influence of the retransmission timer R is linear. In the following studies, we set R to 30 seconds.

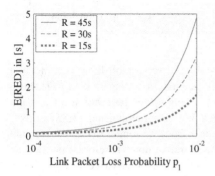

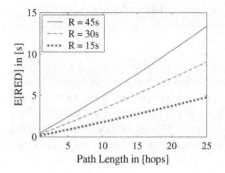

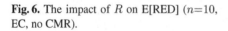

Fig. 6. The impact of R on E[RED] (n=10, EC, no CMR).

Fig. 7. The impact of R on E[RED] (p_l=10^{-2}, EC, no CMR).

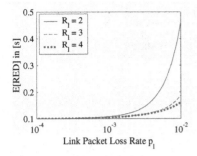
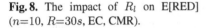

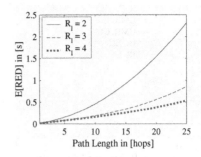

Fig. 8. The impact of R_l on E[RED] ($n=10$, $R=30s$, EC, CMR).

Fig. 9. The impact of R_l on E[RED] ($p_l=10^{-2}$, $R=30s$, EC, CMR).

The CMR option has been standardized [6] to reduce the signaling delay for RSVP in lossy networks. The retransmission interval R_f influences the retransmission times at most linearly and we set it to 0.5 seconds. Figure 8 shows that CMR greatly reduce the response time of the system when we compare this alternative with Figure 6. For $p_l = 10^{-3}$ E[RED] is still negligible. CMR reduce the influence of the refresh interval R and yield shorter RED. Even a single retransmission ($R_l = 2$) reduces E[RED] from about 3 to 0.5 seconds for $p_l = 10^{-2}$. For large loss rates the effect of the rapid retry limit R_l can be well observed. In Figure 9 ($p_l = 10^{-2}$) we observe that E[RED] increases like in Figure 7 about linearly with the length of the reservation path. However, the absolute delay for CMR ($R_l = 3$) is less than 10% compared to the result without CMR. Therefore, CMR is even more important for long paths since the absolute saved delay is larger. In the following experiments, we set $R_l = 3$.

So far, we have considered only EC: When a control message is lost on the way from sender to receiver, the next one is triggered again at the initial node. With CC, the nodes are more active: The last node, that has received an initial PATH or RESV control messages, generates a refresh messages by itself by latest after R time, i. e. the remaining distance to the receiver is shorter than with EC. However, Figure 10 shows that this has no impact: Without CMR, CC is hardly better than EC. In contrast, the effect of CMR is evident: With CMR, E[RED] stays very small while without CMR, E[RED] rises notably. In case of CC we have fewer losses between CMR peers (link loss probabilities) than for EC (end-to-end loss probabilities). Therefore, we can see a difference for high loss rates also between EC and CC with CMR. In case of CMR, E[RED] stays below one second (cf. Figure 11) which means that most of the delay is produced by the link transmission delay ($t_l = 10$ milliseconds) and that the delay due to the refresh interval R is minimized.

3.3 Reservation Teardown Delay

Reserved but unused resources can not be allocated to other connections until their reservation is torn down. The resources are not profitable for that time, so it is important to keep the RTD small. The RTD increases if a TEARDOWN message is lost on the way from the sender to receiver. Due to the soft state concept, an intermediate node tears down the reservation when its cleanup timer expires after L time. If only the endpoints

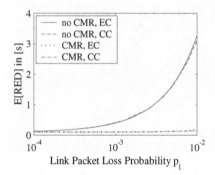
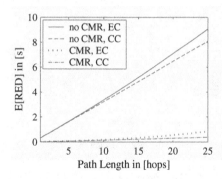

Fig. 10. The impact of the control and CMR option on E[RED] (n=10, R=30s, R_l=3).

Fig. 11. The impact of the control and CMR option on E[RED] (p_l=10^{-2}, R=30s, R_l=3).

control the session, all nodes time out after L since the terminating endpoint refrains from sending refresh messages. If the reservation is under CC (like in RSVP), the second router is not refreshed by the sender but it keeps on generating refresh messages autonomously in periodic intervals. When it times out after L time, its TEARDOWN message can be lost as well. This leads in the worst case to a maximum RTD of $n \cdot L$ for a reservation length of n links.

However, Figure 12 shows that the control option has no impact on E[RTD]. The influence of L dominates E[RTD] for high link loss probabilities by a linear law. The same phenomenon can be observed in Figure 13. Here, it is visible that CC is clearly worse than EC. But the value L for the expiration timer has still greater impact. In the following, we set $L = 3 \cdot R$.

We investigate the influence of the CMR and its parameters on E[RTD]. With CMR, up to R_l TEARDOWN messages are sent repeatedly until an acknowledgement returns. A comparison of Figure 14 and Figure 12 shows that this reduces E[RTD] notably. Especially CC profits from CMR: Their E[RTD] is now shorter than for EC. A rapid

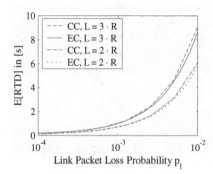
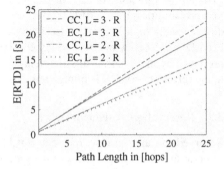

Fig. 12. The impact of L and the control option on E[RTD] (n=10, R=30s, no CMR).

Fig. 13. The impact of L and the control option on E[RTD] (p_l=10^{-2}, R=30s, no CMR).

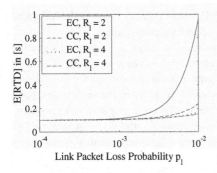

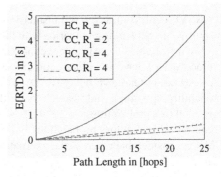

Fig. 14. The impact of the CMR parameter R_l and the control option on E[RTD] (n=10, R=30s, L=3, CMR).

Fig. 15. The impact of the CMR parameter R_l and the control option on E[RTD] (p_l=10^{-2}, R=30s, L=3, CMR).

retry limit of $R_l = 2$ already suffices to keep E[RTD] small, however EC with $R_l = 4$ exhibits an excellent performance as well. Figure 15 shows that a single retransmission reduces E[RTD] for EC to 25% compared to without CMR (cf. Figure 13). But this combination performs still quite poorly in contrast to the other configurations.

3.4 Concluding Remarks

If packet loss rates are high, CMR are a powerful method to reduce the response time in sigaling. CC is an alternative to EC to improve the session stability in that case. In our investigations CMR turned out to be more effective than CC to reduce E[RED] and E[RTD]. The combination of both techniques is possible [6] and leads to optimum results. However, both extensions increase the software complexity of existing RSVP solutions which may have drawbacks that are not respected in this study.

4 Summary

In this paper, we gave an overview of several resource reservation protocols. We presented their basic operations with focus on their information passing concepts.

We explained RSVP and various extensions that reduce the number of refresh messages, make it more robust against packet losses and more scalable for the use in transit networks. Apart from RSVP, light-weight protocols as Boomerang and YESSIR were presented. BGRP is intended for reservation aggregation between autonomous systems. All these approaches derive the possible traffic intensity and the remaining capacity from the stored flow information. Measurement based admission control (MBAC) substitutes this by an estimation of the current network load. We also presented the principle of stateless reservation protocols which is a special MBAC approach.

In RSVP every node supporting a reservation is actively involved in keeping the reservation alive (common control). This is unlike in Boomerang or YESSIR (endpoint control) where only sender and receiver trigger refresh messages. Recently, an option for RSVP control message retransmission (CMR) was created to make RSVP more responsive in networking scenarios with high packet loss probabilities.

We investigated these protocol concepts that are basic features for general signaling protocols. They have an influence on the reservation establishment delay (RED) and on the reservation teardown delay (RTD). Their impact depends both on the packet loss probability of a single link as well as on the number of hops in the reservation path. For small loss probabilities ($< 10^{-3}$), however, their effect is negligible. In networking scenarios with high loss probabilities like wireless networks, the performance gain by CMR is considerable whereas the alternative endpoint or common control plays only a marginal role.

References

1. Chiueh, T., Neogi, A., Stirpe, P.: Performance Analysis of an RSVP-Capable Router. In: 4th Real-Time Technology and Applications Symposium. (1998)
2. Karsten, M., Schmitt, J., Steinmetz, R.: Implementation and Evaluation of the KOM RSVP Engine. In: Infocom, IEEE (2001) 1290–1299
3. Fehér, G., Németh, K., Czslényi, I.: Performance Evaluation Framework for IP Resource Reservation Signalling. In: 8th IFIP Workshop of Performance Modelling and Evaluation of ATM & IP Networks. (2000)
4. Pan, P., Schulzrinne, H.: Processing Overhead Studies in Resource Reservation Protocols. In: 17th International Teletraffic Congress. (2001)
5. Eramo, V., Mocci, U., Fratini, M., Listanti, M.: Reliability Evaluation of RSVP. In: 8th International Telecommunication Network Planning Symposium (Networks 98). (1998)
6. Berger, L., Gan, D.H., Swallow, G., Pan, P., Tommasi, F., Molendini, S.: RFC2961: RSVP Refresh Overhead Reduction Extensions. http://www.ietf.org/rfc/rfc2961.txt (2001)
7. Schulzrinne, H., Rosenberg, J.: Internet Telephony: Architecture and Protocols - an IETF Perspective. Computer Networks **31** (1999) 237–255
8. Braden, B., et al.: RFC2205: Resource ReSerVation Protocol (RSVP) - Version 1 Functional Specification. ftp://ftp.isi.edu/in-notes/rfc2205.txt (1997)
9. Wroclawski, J.:   RFC2210: The Use of RSVP with IETF Integrated Services. ftp://ftp.isi.edu/in-notes/rfc2210.txt (1997)
10. Pan, P., Schulzrinne, H.: Staged Refresh Timers for RSVP. In: 2nd Global Internet Conference. (1997)
11. Baker, F., Iturralde, C., Le Faucheur, F., Davie, B.: Rfc3175: Aggregation of RSVP for IPv4 and IPv6 Reservations. http://www.ietf.org/rfc/rfc3175.txt (2001)
12. Rosen, E.C., Viswanathan, A., Callon, R.: RFC3031: Multiprotocol Label Switching Architecture. http://www.ietf.org/rfc/rfc3031.txt (2001)
13. Menth, M.: A Scalable Protocol Architecture for End-to-End Signaling and Resource Reservation in IP Networks. In: 17th International Teletraffic Congress. (2001) 211–222
14. Menth, M., Hauck, N.: A Graph-Theoretical Concept for LSP Hierarchies. Technical Report, No. 287, University of Würzburg, Institute of Computer Science (2001)
15. Fehér, G., Németh, K., Maliosz, M., Czslényi, I., Bergkvist, J., Ahlard, D., Engborg, T.: Boomerang - A Simple Protocol for Resource Reservation in IP Networks. In: IEEE Workshop on QoS Support for Real-Time Internet Applications, Vancouver, Canada (1999)
16. Pan, P., Schulzrinne, H.: YESSIR: A Simple Reservation Mechanism for the Internet. Computer Communication Review **29** (1999)
17. Schulzrinne, H., Casner, S., Frederick, R., Jacobson, V.: RFC1889: RTP - A Transport Protocol for Real-Time Applications. ftp://ftp.isi.edu/in-notes/rfc1889.txt (1996)

18. Pan, P., Schulzrinne, H.: BGRP: A Tree-Based Aggregation Protocol for Inter-domain Reservations. Journal of Communications and Networks **2** (2000) 157–167
19. Breslau, L., Knightly, E.W., Shenker, S., Zhang, H.: Endpoint Admission Control: Architectural Issues and Performance. In: ACM SIGCOMM. (2000)
20. Breslau, L., Jamin, S., Shenker, S.: Comments on the Performance of Measurement-Based Admission Control Algorithms. In: Infocom. (2000) 1233–1242
21. Más, I., Karlsson, G.: PBAC: Probe-Based Admission Control. In: 2nd International Workshop on Quality of future Internet Services (QofIS2001). (2001)
22. Stoica, I., Zhang, H.: Providing Guaranteed Services Without Per Flow Management. Computer Communication Review **29** (1999)
23. Almesberger, W., Ferrari, T., Le Boudec, J.Y.: SRP: a Scalable Resource Reservation for the Internet. In: IFIP 6th International Workshop on Quality of Service (IWQoS'98). (1998)
24. Eriksson, A., Ghermann, C.: Robust and Secure Light-weight Resource Reservation of Unicast IP Traffic. In: IFIP 6th International Workshop on Quality of Service (IWQoS'98). (1998)

Handling Multiple Bottlenecks in Web Servers Using Adaptive Inbound Controls

Thiemo Voigt[1] and Per Gunningberg[2]

[1] SICS, Box 1263, SE-16429 Kista, Sweden,
thiemo@sics.se
[2] Uppsala University, Box 337, SE-75105 Uppsala, Sweden,
Per.Gunningberg@it.uu.se

Abstract. Web servers become overloaded when one or several server resources are overutilized. In this paper we present an adaptive architecture that prevents resource overutilization in web servers by performing admission control based on application-level information found in HTTP headers and knowledge about resource consumption of requests. In addition, we use an efficient early discard mechanism that consumes only a small amount of resources when rejecting requests. This mechanism first comes into play when the request rate is very high in order to avoid making uninformed request rejections that might abort ongoing sessions. We present our dual admission control architecture and various experiments that show that it can sustain high throughput and low response times even during high load.

1 Introduction

Web servers need to be protected from overload because web server overload can lead to high response times, low throughput and potentially loss of service. Therefore, there is a need for efficient admission control to maintain high throughput and low response time during periods of peak server load. Servers become overloaded when one or several critical resources are overutilized and become the bottleneck of the server system. The main server resources are the network interface, CPU and disk [8]. Any of these may become the server's bottleneck, depending on the kind of workload the server is experiencing [10]. For example, the majority of CPU load is caused by a few CGI requests [4]. The network interface typically becomes overutilized when the server concurrently transmits several large files.

In this paper, we report on an adaptive admission control architecture that utilizes the information found in the HTTP header of incoming requests. Combining this information with knowledge about the resource consumption we can avoid resource overutilization and server overload. We call our approach resource-based admission control.

The current version of our admission control architecture is targeted at overloaded single node web servers or the back-end servers in a web server cluster. The load on web servers can be reduced by distributed web server architectures

G. Carle and M. Zitterbart (Eds.): PfHSN 2002, LNCS 2334, pp. 50–68, 2002.

that distribute client requests among multiple geographically dispersed server nodes in a user-transparent way [29]. Another approach is to redirect requests to web caches. For example, in the distributed cache system Cachemesh [30] each cache server becomes the designated cache for several web sites. Requests for objects not in the cache are forwarded to the responsible cache. However, not all web data is cacheable, in particular dynamic and personalized data. Also, load balancing mechanisms on the front-ends of web server clusters can help to avoid overload situations on the back-end servers. However, sophisticated load-balancing cannot replace proper overload control [13]. Another existing solution to alleviate the load on web servers based on a multi-process architecture such as Apache is to limit the maximum·number of server processes. However, this approach limits the number of requests that the server can handle concurrently and can lead to performance degradation.

The main contribution of this work is an adaptive admission control architecture that handles multiple bottlenecks in server systems. Furthermore, we show how we can use TCP SYN policing and HTTP header-based control in a combined way to perform efficient and yet informed web server admission control.

Resource-based admission control uses a kernel-based mechanism for overload protection and service differentiation called *HTTP header-based connection control* [31]. HTTP header-based connection control allows us to perform admission based on application-level information such as URL, sets of URLs (identified by, for example, a common prefix), type of request (static or dynamic) and cookies. HTTP header-based control uses token bucket policers for admission control. HTTP header-based connection control is used in conjunction with filter rules that specify application-level attributes and the parameters for the associated control mechanism, i.e. the rate and bucket size of the policer.

Our idea to avoid overutilization of server resources is the following: we collect all objects that when requested are the main consumers of the same server resource into one directory. Thus, we have one directory for each critical resource. Each of these directories is then moved into a separate part of the web server's directory tree. We associate a filter rule with each of these directories. Hence, we can use HTTP header-based control to protect each of the critical resources from becoming overutilized. For example, CPU-intensive requests reside in the `cgi-bin` directory and a filter rule specifying the application-level information (URL prefix `/cgi-bin`) is associated with the content at this location.

When the request rate reaches above a certain level, resource-based admission control cannot prevent overload, for example during flash crowds. When such situations arise, we use *TCP SYN policing* [31]. This mechanism is efficient in terms of resource consumption of rejected requests because is provides "early discard". The admission of connection requests is based on network-level attributes such as IP addresses and a token bucket policer.

This paper also presents novel mechanisms that dynamically set the rate of the token bucket policers based on the utilization of the critical resources. Since the web server workload frequently changes, for example when the popularity of documents or services changes, assigning static rates that work under these changing conditions may either lead to underutilization of the system when the rates are too low or there is a risk for overload when the rates are too high. The adaptation of the rates is done using feedback control loops. Techniques from control theory have been used successfully in server systems before [2,14,11,20].

We have implemented this admission control architecture in the Linux OS and using an unmodified Apache web server, we conducted experiments in a controlled network. Our experiments show that overload protection and adaptation of the rates works as expected. Our results show higher throughput and much lower response times during overload compared to a standard Apache on Linux configuration.

The rest of the paper is structured as follows. Section 2 presents the system architecture including the controllers. Section 3 presents experiments that evaluate various aspects of our system. Section 4 discusses architectural extensions. Before concluding, we present related work in Section 5.

2   Architecture

2.1   Basic Architecture

Our basic architecture deploys mechanisms for overload protection and service differentiation in web servers that have been presented earlier [31]. These mechanisms control the amount and rate of work entering the system (see Figure 1).

TCP SYN policing limits acceptance of new SYN packets based on compliance with a token bucket policer. Token buckets have a token rate, which denotes the average number of requests accepted per second and a bucket size which denotes the maximum number of requests accepted at one time. TCP SYN policing enables service differentiation based on information in the TCP and IP headers of the incoming connection request (i.e, the source and destination addresses and port numbers).

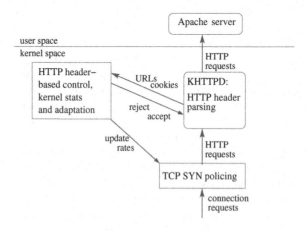

Fig. 1. Kernel-based architecture

HTTP header-based connection control is activated when the HTTP header is received. Using this mechanism a more informed control is possible which provides the ability to, for example, specify lower access rates for CGI requests than other requests that are less CPU-intensive. This is done using filter rules, e.g. checking URL, name and type.

The architecture consists of an unmodified Apache server, the TCP SYN policer, the in-kernel HTTP GET engine khttpd [24] and a control module that implements HTTP header-based control, monitors critical resources and adapts the acceptance rates. Khttpd is used for header parsing only. After parsing the request header it passes the URLs and cookies to the control module that performs HTTP header-based control. If the request is rejected, khttpd resets the connection. In our current implementation, this is done by sending a TCP RST back to the client. If the request is accepted it is added to the Apache web server's listen queue. TCP SYN policing drops non-compliant TCP SYNs. This implies that the client will time-out waiting for the SYN ACK and retry with an exponentially increasing time-out value. For a more detailed discussion, see [31].

Both mechanisms are located in the kernel which avoids the context switch to user space for rejected requests. The kernel mechanisms have proven to be much more efficient and scalable than the same mechanisms implemented in the web server [31].

2.2 The Dual Admission Control Architecture

Our dual admission control architecture is depicted in Figure 2. In the right part of the figure we see the web server and some of its critical resources. With each of these resources, a filter rule and a token bucket policer is associated to avoid overutilization of the resource, i.e. we use the HTTP header-based connection control mechanism. For example, a filter rule /cgi-bin and an associated token bucket policer restrict the acceptance of CPU-intensive requests. On receipt of a request, the HTTP header is parsed and matched against the filter rules. If there is a match, the corresponding token bucket is checked for compliance. Compliant

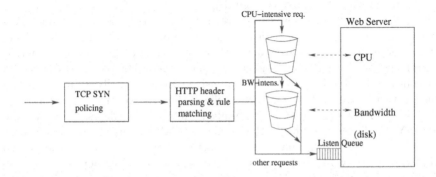

Fig. 2. Admission control architecture

requests are inserted into the listen queue. We call this part of our admission control architecture resource-based admission control.

For each of the critical resources, we use a feedback control loop to adapt the token rate at which we accept requests in order to avoid overutilization of the request. We do not adapt the bucket size but assume it to be fixed. Note, that the choice of the policer's bucket size is a trade-off [31]. When we have a large bucket size, malicious clients can send a burst to put high load on the machine, whereas when the bucket size is small, clients must come at regular intervals to make full use of the token rate. Furthermore, when the bucket size is smaller than the number of parallel connections in a HTTP 1.0 browser, a client might not be able to retrieve all the embedded objects in a HTML-page since requests for these objects usually come in a burst after the client has received the initial HTML page.

Note that we do not perform resource-based admission control on all requests. Requests such as those for small static files do not put significant load on one resource. However, if requested at a sufficiently high rate, these requests can still cause server overload. Hence, admission control for these requests is needed. We could insert a default rule and use another token bucket for these requests. Instead, we have decided to use TCP SYN policing and police all incoming requests. The main reason for this is TCP SYN policing's early discard capability. Also for TCP SYN policing, we adapt the token rate and keep the bucket size fixed.

One of our design goals for the adaptation mechanisms is to keep TCP SYN policing inactive while resource-based admission control can protect resources from being overutilized. When performing resource-based admission control, the whole HTTP header has been received and can be checked not only for URLs but also for other application-level information, such as cookies. This gives us the ability to identify ongoing sessions or premium customers. TCP SYN policing's admission control is based on network-level information only and cannot assess such application-level information. Note that this does not mean that TCP SYN policing is not worthwhile [31]. Firstly, TCP SYN policing can provide service differentiation based on network-level attributes. Secondly, TCP SYN policing is more efficient than HTTP header-based control in the sense that less resources are consumed when a request is discarded.

Our architecture uses several control loops to adapt the rate of the token bucket policers: One for each critical resource and one to adapt the SYN policing rate. A consequence of this approach is that the interaction between the different control loops might cause oscillations. Fortunately, requests to large static files do not consume much CPU while CPU-intensive requests usually do not consume much network bandwidth. Thus, we believe that the control loops for these resources will not experience any significant interaction effects. Adapting the TCP SYN policing rate affects the number of CPU-intensive and bandwidth-intensive requests, which may cause interactions between the control loops. To avoid this effect, we increase rates quite conservatively. Furthermore, in none of our experiments we have seen an indication that such an interaction might actually occur. One of the reasons for this is that TCP SYN policing becomes active when the acceptance rate of CPU-intensive requests is very low, i.e. most of the CPU-intensive requests are discarded.

2.3   The Control Architecture

Our control architecture is depicted in Figure 3. The monitor's task is to monitor the utilization of each critical resource and pass it to the controller. The controllers adapt the rates for the admission control mechanisms. We use one controller for the CPU utilization and one for the bandwidth on the outgoing network interface. We call the former CPU controller and the latter bandwidth controller. Both high CPU utilization and dropped packets on the networking interface can lead to long delays and low throughput. Other resources that could be controlled are disk I/O bandwidth and memory. In addition, we use a third controller that is not responsible for a specific resource but performs admission control on all requests, including those that are not associated with a specific resource. The latter controller, called SYN controller, controls the rate of the TCP SYN policer.

Since different resources have different properties, we cannot use the same controller for each resource. The simplest resource to control is the CPU. The CPU utilization changes directly with the rate of CPU-intensive requests. This makes it possible to use a proportional (P) controller. The equation that computes the new rates is called the *control law*. For our P-controller the control law is:

$$rate_{cgi}(t+1) = rate_{cgi}(t) + K_{P\_CPU} * e(t) \tag{1}$$

where $e(t) = CPU\_util_{ref} - CPU\_util(t)$, i.e. the difference between the *reference* or desired CPU utilization and the current, measured CPU utilization. $rate_{cgi}(t)$ is the acceptance rate for CGI-scripts at time t. K_{P_CPU} is called the *proportional*

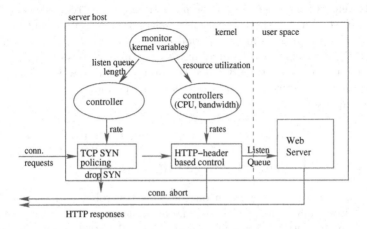

Fig. 3. The control architecture

gain. It determines how fast the system will adapt to changes in the workload. For higher $K_{\text{P\_CPU}}$ the adaptation is faster but the system is less stable and may experience oscillations [12].

The other two controllers in our architecture base their control laws on the length of queues: The SYN controller on the length of the listen queue and the bandwidth controller on the length of the queue to the network interface. The significant aspect here is actually the change in the queue length. This derivative reacts faster than a proportional factor. This fast reaction is more crucial for these controllers since the delay between the acceptance decision and the actual occurrence of high resource utilization is higher than when controlling CPU utilization. For example, the delay between accepting too many large requests and overflow of the queue to the network interface is non-negligible. One reason for this is that it takes several round-trip times until the TCP congestion window is sufficiently large to contribute to overflow of the queue to the network interface.

We decided therefore to use a proportional derivative (PD) controller for these two controllers. The derivative is approximated by the difference between the current queue length and the previous one, divided by the number of samples. The control law for our PD-controllers is:

$$rate(t+1) = rate(t) + K_{\text{P\_Q}} * e(t) + K_{\text{D\_Q}} * (queue\_len(t) - queue\_len(t-1)) \quad (2)$$

where $e(t) = queue\_len_{\text{ref}} - queue\_len(t)$. The division is embedded in $K_{\text{D\_Q}}$. $K_{\text{D\_Q}}$ is the *derivative gain*.

We imposed some conditions on the equations above. Naive application of Equation 1 results in an increase in the acceptance rate when the measured value is below the reference value, i.e. the resource could be utilized more. However, it is not meaningful to increase the acceptance rate, when the filter rule has less hits than the specified token rate. For example, if we allow 50 CGI requests/sec, the current CPU utilization is 60%, the reference value for CPU utilization is 90% and the server has received 30 CGI requests during the last second, it does not make sense to increase the rate to more than 50 CGI requests/sec. On the contrary, if we increase the rate in such a situation, we would end up with a very high acceptance rate after a period of low server load. Hence, when the measured CPU utilization is lower than the reference value, we have decided to update the acceptance rate only when the number of hits was at least 90% of the acceptance rate during the previous sampling period.

Thus, Equation 1 rewrites as:

$$rate_{\text{cgi}}(t+1) = \begin{cases} rate_{\text{cgi}}(t) & \text{if } (\# \text{ hits }) < 0.9 * rate_{\text{cgi}}(t) \wedge \\ & \quad CPU\_util(t) < CPU\_util_{\text{ref}} \\ rate_{\text{cgi}}(t) + K_{\text{P\_CPU}} * e(t) & \text{otherwise} \end{cases}$$

We impose a similar condition on the SYN controller and the bandwidth controller. Both adapt the rates based the queue lengths. The listen queue and the queue to the network interface usually have a length of zero. This means, that the length of the queue is below the reference value. For the same reason as in the discussion above, if the queue length is below the reference value, we update the acceptance rate only when the length of the queue has changed.

When performing resource-based admission control, we do not police all requests, even if all requests consume resources at least some CPU. Thus, if the CPU utilization is already high, i.e. it is above the reference value, we do not want to increase the amount of work that enters the system since this might cause server overload. Thus, we increase the TCP SYN policing rate only when the CPU utilization is below the reference value.

An important decision is the choice of the sampling rate. For ease of implementation, we started with a sampling rate of one second. To obtain the current CPU utilization, we can use the so-called *jiffies* that the Linux kernel provides. Jiffies measure the time the CPU has spent in user, kernel and idle mode. Since 100 jiffies are equivalent to one second, it is trivial to compute the CPU utilization during the last second. Since even slow web servers can process several hundred requests per second, a sampling rate of one second might be considered long. The question is if we do not miss important events such as the listen queue filling up. This would be the case given that the requests entering the server during one second was not limited. However, TCP SYN policing limits the number of requests entering the system. This bounds the system state changes between sampling points and allows us to use a sampling rate of one second. This is an acceptable solution since the experiment in Section 3.3 shows that the control mechanisms still adapt quickly when we expose the server to sudden high load.

The number of packets queued on an outgoing interface can change quite rapidly. When the queue is full, packets have to be dropped. The TCP connections that experience drops back off and thus less packets are inserted into the queue. We have observed that the queue length has changed from maximum length to zero and back to maximum length within 20 milliseconds. To avoid incorporating such an effect when computing new rates, we sample the queue length to that interface more frequently and compute an average every second. Using this average, the controller updates the rates every second.

3   Experiments

Our testbed consists of a server and two traffic generators connected via a 100 Mb/sec Ethernet switch. The server machine is a 600 MHz Athlon with 128 MBytes of memory running Linux 2.4. The traffic generators run on a 450 MHz Athlon and a 200 MHz Pentium Pro. The server is an unmodified Apache web server, v.1.3.9., with the default configuration, i.e. a maximum of 150 worker processes. We have extended the length of the web server's listen queue from 128 to 1024. Banga and Druschel [5] argue that a short listen queue can limit the throughput of web servers. A longer queue will not have an effect on the major results.

Parameter Settings

We have used the following values for the control algorithms: The reference values for the queues are set to 100 for the listen queue and 35 for the queue to the network interface which has a length of 100. These values are chosen arbitrarily but they can be chosen lower without significant impact on the stability since the queue lengths are mostly zero. Repeating for example the experiment in Section 3.3 with reference values larger than 20 for the listen queue length leads to the same results. The reference value for CPU utilization is 90%. We chose this value since it allows us to be quite close to the maximum utilization while higher values would more often lead to 100% CPU utilization during one sampling period.

The proportional gain for the CPU-controller is set to 1/5. We have obtained this value by experimentation. In our experiments we saw that for gains larger than 1/4, the CGI acceptance rate oscillates between high and low values, while it is stable for smaller values. The proportional gain for the SYN and bandwidth controller is set to 1/16, the derivative gain to 1/4. These values were also obtained by experimentation. When both gains are larger than 1/2, the system is not stable, i.e. the change in both the length of the listen queue and the rates is high when the server experiences high load. When the derivative gain is higher than the proportional gain, the system reacts fast to changes in the queue lengths and the queues rarely grow large when starting to grow. We consider these values to be specific to the server machine we are using[1]. However, we expect them to hold for all kinds of web workloads for this server since we use a realistic workload as described in the next section. The bucket size of the token bucket used for TCP SYN policing is set to 20 unless explicitly mentioned. The token buckets for HTTP header-based controls have a bucket size of five.

3.1   Workload

For client load generation we use the sclient traffic generator [5]. Sclient is able to generate client request rates that exceed the capacity of the web server. This is done by aborting requests that do not establish a connection to the server in a specified amount of time. Sclient in its unmodified version requests a single file. For most of our experiments we have modified sclient to request files according to a workload that is derived from the surge traffic generator [6]:

1. The size of the files stored on the server follows a heavy tailed distribution.
2. The request size distribution is heavy tailed.
3. The distribution of popularity of files follows Zipf's Law. Zipf's Law states that if the files are ordered from most popular to least popular, then the number of references to a file tends to be inversely proportional to its rank.

Determining the total number of requests for each file on the server is also done using surge. We separated the files in two directories on the server. The files larger

---

[1] The values also worked well on another machine we tested, but we do not assume this is the general case.

than 50 KBytes were put into one directory (`/islarge`), the smaller files into another directory. Harchol-Balter et al. [23] divide static files into priority classes according to size and assign files larger than 50 KBytes into the group of largest files. We made 20% of the requests for small files dynamic. The dynamic files used in our experiments are minor modifications of standard Webstone [15] CGI files and return a file containing randomly generated characters of the specified size. The fraction of dynamic requests varies from site to site with some sites experiencing more than 25% dynamic requests [22,21]. For the acceptance rate of both CGI-scripts and large files, minimum rates can be specified. The reason for this is that the processing of CGI-scripts or large files should not be completely prevented even under heavy load. This minimum rate is set to 10 reqs/sec in our experiments.

In the next sections we report on the following experiments: In the first experiment we show that the combination of resource-based admission control and TCP SYN policing works and adapts the rates as expected. In this experiment the CPU is the major bottleneck. In the following experiment, we expose the system to a sudden high load and study the behaviour of the adaptation mechanisms under such circumstances. In the experiment in Section 3.4, we make the bandwidth on the interface a bottleneck and show how resource-based admission control can prevent high response times and low throughput. In the last experiment, we show that the adaptation mechanisms can cope with more bursty request arrival distributions.

3.2   CPU Time and Listen Queue Length

In this experiment, we use two controllers: the CPU controller that adapts the acceptance rate of CGI-scripts and the SYN controller. As mentioned earlier, the reference for adapting the rate of CGI-scripts is the CPU utilization and the reference for the TCP SYN policing rate is the listen queue length. About 20% of the requests are for dynamic files (CGI-scripts). In the experiment, we vary the request rate across runs. The goals of the experiment are: (i) show that the control algorithms and in particular resource-based admission control prevent overload and sustain high throughput and low response time even during high load; (ii) show that TCP SYN policing becomes active when resource-based admission control alone cannot prevent server overload; (iii) show that the system achieves high throughput and low response times over a broad range of possible request rates.

For low rates, we expect that no requests should be discarded. When the request rate increases, we expect that the CPU becomes overutilized mostly due to the CPU-intensive CGI-scripts. Hence, for some medium request rates, policing of CGI-scripts is sufficient and TCP SYN policing will not be active. However, when the offered load increases beyond a certain level, the processing capacity of the server will not be able keep up with the request rate even when discarding most of the CPU-intensive requests. At that point, the listen queue will build up and thus TCP SYN policing will become active.

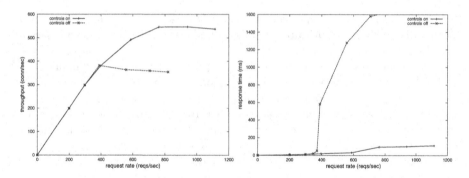

Fig. 4. Comparison standard system (no controls) and system with adaptive overload control

Figure 4 compares the throughput and response times for different request rates. When the request rate is about 375 reqs/sec, the average response time increases and the throughput decreases when no controls are applied[2]. Since our workload contains CPU-intensive CGI-scripts, the CPU becomes overutilized and cannot process requests with the same rate as they arrive. Hence, the listen queue builds up which contributes additionally to the increase of the response time.

Using resource-based admission control, the acceptance rate of CGI-scripts is decreased which prevents the CPU from becoming a bottleneck and hence keeps the response time low. Decreasing the acceptance rate of CGI-scripts is sufficient until the request rate is about 675 reqs/sec. At this point the CGI acceptance rate reaches the predefined minimum and cannot be decreased anymore despite the CPU utilization being greater than the reference value. As the server's processing rate is lower than the request rate, the listen queue starts building up. Due to the increase of the listen queue, the controller computes a lower TCP SYN policing rate which limits the number of accepted requests. This can be seen in the left-hand graph where the throughput does not increase anymore for request rates higher than 800 reqs/sec. The right-hand graph shows that the average response time increases slightly when TCP SYN policing is active. Part of this increase is due to the additional waiting time in the listen queue.

We have repeated this experiment with workloads containing 10% dynamic requests and only static requests. If more requests are discarded using HTTP header-based control, the onset of TCP SYN policing should happen with higher request rates. The results in Table 1 show that this is indeed the case. When the fraction of dynamic requests is 20%, TCP SYN policing sets in at about 675 reqs/sec while the onset for SYN policing is at about 610 reqs/sec when all requests are for static files.

---

[2] For higher request rates than those shown in the graph the traffic generator runs out of socket buffers when no controls are applied.

Table 1. Request rate for which SYN policing becomes active for different fractions of dynamic requests in the workload

fraction dynamic reqs. (%)	req rate for onset of SYN policing (reqs/sec)
20	675
10	640
0	610

In summary, for low request rates, we prevent server overload using resource-based admission control that avoids over-utilization of the resource bottleneck, in this case CPU. For high request rates, when resource-based admission control is not sufficient, TCP SYN policing reduces the overall acceptance rate in order to keep the throughput high and response times low.

3.3   Exposure to a Very High Load

In this experiment we expose the server to a sudden high load and study the behaviour of the control algorithms. Such a load exposure could occur during a flash crowd or a Denial-of-Service (DoS) attack. We start with a relatively low request rate of 300 reqs/sec. After 50 seconds we increase the offered load to 850 reqs/sec and sustain this high request rate for 20 seconds before we decrease it to 300 reqs/sec again. We set the initial TCP SYN policing rate to 1000 reqs/sec.

Figure 5 shows that the TCP SYN policing rate decreases very quickly when the request rate is increased at time 50. This rapid decrease is caused by both parts of the control algorithms in Equation 2. First, since the length of the listen queue increases quickly, the contribution of the derivative part is high. Second, the absolute length of the listen queue is at that time higher than the reference value. Thus, the contribution of the proportional part of Equation 2 is high as well. The TCP SYN policing rate does not increase to 1000 again after the period of high load. However, we can see that around time 70, the policing rate increases to around 340 which is sufficiently high so that no requests need to be discarded

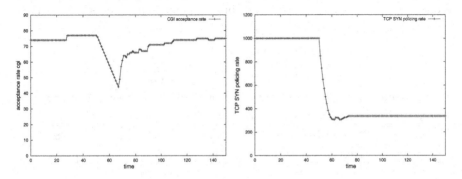

Fig. 5. Adaptation under high load (left CGI acceptance rate, right SYN policing rate)

by the SYN policer when the request rate is 300 reqs/sec. For higher request rates after the period of high load, the SYN policing rate settles at higher rates.

As expected, the CGI acceptance rate does not decrease as fast. With K_{P_CPU} being 1/5, the decrease of the rate is at most two per sampling point during the period of high load. Figure 5 also shows that the CGI-acceptance rate is restored fast after the period of high load. At a request rate of 300 reqs/sec, the CPU utilization is between 70 and 80%. Thus, the absolute difference to the reference value is larger than during the period of high load which enables faster increase than decrease of the CGI acceptance rate. At time 30 in the left-hand graph, we can see the CGI acceptance rate jump from 74 to 77. The reason for this jump is that during the last sampling period, the number of hits for the corresponding filter rule was above 90%, while it was otherwise below 90% until time 50.

3.4   Outgoing Bandwidth

Despite the fact that the workload used in the previous section contains some very large files, there were very few packet drops on the outgoing network interface. In the experiments in this section we make the bandwidth of the outgoing interface a bottleneck by requesting a large static file of size 142 KBytes from another host. The original host still requests the surge-like workload at a rate of 300 reqs/sec. From Figure 4 in Section 3.2, we can see that the server can cope with the workload from this particular host requested at this rate. The request of the large static file will cause overutilization of the interface and a proportional drop of packets to the original host.

Without admission control, we expect that packet drops on the outgoing interface will cause lower throughput and in particular higher average response times by causing TCP to back off due to the dropped packets. We therefore insert a rule that controls the rate at which large files are accepted. Large files are identified by a common prefix (`/islarge`). The aim of the experiment is to show that by adapting the rate with that requests for large files are accepted, we can avoid packets drops on the outgoing interface.

We generate requests to the large file with a rate of 50 and 80 reqs/sec. The results are shown in Table 2. As expected the response times for both workloads become very high when no controls are applied. In our experiments, we observed

Table 2. Outgoing bandwidth

req rate large workload	metric	workload			
		large workload		surge workload	
		no controls	controls	no controls	controls
50 reqs/s	tput (reqs/sec)	46.8	41.5	270.7	289.2
50 reqs/s	response time (ms)	2144	80.5	1394.8	26.9
80 reqs/s	tput (reqs/sec)	55.5	45.8	205.2	285.1
80 reqs/s	response time (ms)	5400	94	3454.5	29.3

that the length of the queue to the interface was always around the maximum value which indicates a lot of packet drops. By discarding a fraction of the requests for large files our controls keep the response time low by avoiding drops in the queue to the network interface. Although the throughput for the large workload is higher when no controls are applied, the sum of the throughput for both workloads is higher using the controls. Note, that when controls are applied the sum of the throughput for both workloads is the same for both request rates (about 331 reqs/sec).

3.5 Burstier Arrival Requests

The sclient program generates web server requests at a constant rate. The resulting requests also arrive at a constant rate to the web server. We have modified the sclient program to generate requests following a Poisson distribution with a given mean. To verify that our controllers can cope with burstier traffic we have repeated the experiment in Section 3.2 with a Poisson distribution. In Figure 6 we show the throughput at constant rate and at a Poisson distribution. The x-axis denotes the mean of the Poisson distribution or the constant request rate of the standard sclient program while the y-axis denotes the throughput. The difference between the two graphs is that the bucket size of the policers is 20 in the left-hand graph and five in the right-hand graph. Traffic generated at a constant rate should be almost independent of the bucket size since it arrives regularly at the server and a new token should always be available given the token rate is sufficient.

The left-hand graph shows that we achieve about the same throughput independent of the distribution of the requests when the policer's bucket size is 20. If the policer's bucket size is small as in the right-hand graph, more requests than necessary are rejected when the distribution of the requests' arrival times is burstier. This experiment shows that our adaptation mechanisms should be able to cope with bursty traffic provided we make a sensible choice of the bucket size.

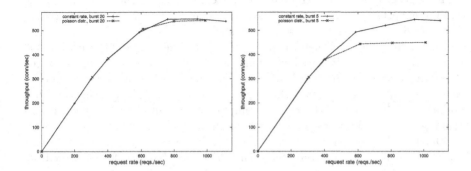

Fig. 6. Comparison between constant request rate and traffic generated according to Poisson distribution for different bucket sizes (left 20, right 5)

4   Architectural Extensions

Our current implementation is targeted towards single node servers or the back-end servers in a web server cluster. We believe that the architecture can easily be extended to LAN-based web server clusters and enhance sophisticated request distribution schemes such as HACC [27] and LARD [26]. In LARD and HACC, the front-end distributes requests based on locality of reference to improve cache hit rates and thus increase performance. Aron et al. [25] increase the scalability of this approach by performing request distribution in the back-ends. In our extended architecture, the front-end performs resource-based admission control. The back-end servers monitor the utilization of each critical resource and propagate the values to the front-end. Based on these values, the front-end updates the rates for the token bucket based policers using the algorithms presented in Section 2.3. After the original distribution scheme has selected the node that is to handle the request, compliance with the corresponding token bucket ensures that critical resources on the back-ends are not overutilized. This way, we consider the utilization of individual resources as a distribution criteria which neither HACC nor LARD do. HACC explicitly combines these performance metrics into a single load indicator.

The front-end also computes the rate of the SYN policers for each back-end based on the listen queue lengths reported by the back-ends. Using these values, the front-end itself performs SYN policing, using the sum of the acceptance rates of the back-ends as acceptance rate to the whole cluster. There are two potential problems: First, the need to propagate the values from the back-ends to the front-end causes some additional delay. If the evaluation of the system shows that this is indeed a problem, we should be able to overcome it by setting more conservative reference values or by increasing the sampling rate. Second, there is a potential scalability problem caused by the need for $n*c$ token buckets on the front-end, where n is the number of back-ends and c the number of critical resources. However, we believe that this is not a significant problem since a token bucket can be implemented by reading the clock (which in kernel space is equal to reading a global variable) and performing some arithmetical operations.

For a geographically distributed web cluster resource utilization of the servers and the expected resource utilization of the requests can be taken into account when deciding on where to forward requests.

Our architecture is implemented as a kernel module, but could be deployed in user space or in a middleware layer. Since our basic architecture is implemented as a kernel module, we have decided to put the control loops in the kernel module as well. An advantage of having the control mechanisms in the kernel is that they are actually executed at the correct sampling rate. But the same mechanisms could be deployed in user space or in a middleware layer.

Our kernel module is not part of the TCP/IP stack which makes it easy to port the mechanisms. The only requirements are availability of timing facilities to ensure correct sampling rates and facilities to monitor resource utilization.

It is also straightforward to extend the architecture to handle persistent connections. Persistent connections represent a challenging problem for web server admission control since the HTTP header of the first request does not reveal any information about the resource consumption of the requests that might follow on the same connection. A solution to this problem is proposed by Voigt and Gunningberg [32] where under server overload persistent connections that are not regarded as important are aborted. The importance is determined by the cookies. This solution can easily be adapted to fit our architecture. If a resource is overutilized, we abort non-important persistent connections with a request matching the filter rule associated with that resource.

It would be interesting to perform studies on user perception. Since the TCP connection between server and client is already set up when HTTP header-based control decides on accepting a request, we can inform the client (in this case the user) by sending a "server busy" notification. TCP SYN policing, on the other hand, just drops TCP SYNs which with current browsers does not provide timely feedback to the client. This is another reason for keeping TCP SYN policing inactive as long as resource-based admission control can prevent server overload.

The netfilter framework which is part of the Linux kernel contains functionality similar to TCP SYN policing. We plan to invest if TCP SYN policing can be reimplemented using netfilter functionality. Other operating systems such as FreeBSD contain firewall facilities that could be used to limit the bandwidth to a web server. It is possible to use such facilities instead of SYN policing. However, since there is no one-to-one mapping between bandwidth and requests, it is harder to control the actual amount of requests entering the web server.

The proposed solution of grouping the objects according to resource demand in the web server's directory tree, is not intuitive and awkward for the system administrator. We assume that this process can be automated using scripts.

5   Related Work

Casalicchio and Colajanni [8] have developed a dispatching algorithm for web clusters that classifies client requests based on their impact on server resources. By dispatching requests appropriately they ensure that the utilization of the individual resources is spread evenly among the server back-ends. Our and their approach have in common that they utilize the expected resource consumption of web requests, however, for different purposes.

Several others have adopted approaches from control theory for server systems. Abdelzaher and Lu [2] use a control loop to avoid server overload and meet individual deadlines for all served requests. They express server utilization as a function of the served rate and the delivered bandwidth [1]. Their control task is to keep the server utilization at $ln2$ in order to guarantee that all deadlines can be met. In our approach we aim for higher utilization and throughput. Furthermore, our approach also handles dynamic requests. In another paper, Lu et al. [14] use a feedback control approach for guaranteeing relative delays in web servers. Parekh et al. [11] use a control-theoretic approach to regulate the maximum number of users accessing a Lotus Notes server. While a focus of these papers is to use control theory to avoid the absence of oscillations, Bhoj et al. [20] in a similar way as we, use a simple controller to ensure that the occupancy of the priority queue of a web server stays at or below a pre-specified target value. Reumann et al. [19] use a mechanism similar to TCP SYN policing to avoid server overload.

Several research efforts have focused on overload control and service differentiation in web servers [3,7,13,9]. *WebQoS* [7] is a middleware layer that provides service differentiation and admission control. Since it is deployed in middleware, it is less efficient compared to kernel-based mechanisms. Cherkasova et al. [9] present an enhanced web server that provides session-based admission control to ensure that longer sessions are completed. Their scheme is not adaptive and rejects new requests when the CPU utilization of the server exceeds a certain threshold. The focus of cluster reserves [28] is to provide performance isolation in cluster-based web servers by managing resources, in their work CPU. Their resource management and distribution strategies do not consider multiple resources.

There are some commercial approaches that deserve mention. Cisco's *LocalDirector* [18] enables load balancing across multiple servers with per-flow rate limits. Inktomi's *Traffic Server C-Class* [17] provides system server overload detection and throttling from traffic spikes and DoS attacks by redistributing requests to caches. Alteon's *Web OS Traffic Control Software* [16] parses HTTP headers to perform URL-based load balancing and redirect requests based on content type to servers.

6   Conclusions

We have presented an adaptive server overload protection architecture for web servers. Using the application-level information in the HTTP header of the requests combined with knowledge about resource consumption of resource-intensive requests, the system adapts the rates at which requests are accepted. The architecture combines the use of such resource-based admission control with TCP SYN policing. TCP SYN policing first comes into play when the load on the server is very high since it wastes less resources when rejecting requests. Our experiments have shown that the acceptance rates are adapted as expected. Our system sustains high throughput and low response times even under high load.

Acknowledgements

This work builds on the architecture that has been developed at IBM TJ Watson together with Renu Tewari, Ashish Mehra and Douglas Freimuth [31]. The authors also want to thank Martin Sanfridson and Jakob Carlström for discussions on the control algorithms and Andy Bavier, Ian Marsh, Arnold Pears and Bengt Ahlgren for valuable comments on earlier drafts of this paper.

This work is partially funded by the national Swedish Real-Time Systems research initiative ARTES (www.artes.uu.se), supported by the Swedish Foundation for Strategic Research.

References

1. Abdelzaher T., Bhatti N.: Web Server QoS Management by Adaptive Content Delivery. Int. Workshop on Quality of Service (1999)
2. Abdelzaher T.,Lu C.: Modeling and Performance Control of Internet Servers. Invited paper, IEEE Conference on Decision and Control (2000)
3. Almeida J., Dabu M., Manikutty A., Cao P.: Providing Differentiated Levels of Service in Web Content Hosting. Internet Server Performance Workshop (1999)
4. Arlitt M., Williamson C.: Web Server Workload Characterization: The Search for Invariants. Proc. of ACM Sigmetrics (1996)
5. Banga G., Druschel P.: Measuring the Capacity of a Web Server. USENIX Symposium on Internet Technologies and Systems (1997)
6. Barford P., Crovella M.: Generating Representative Web Workloads for Network and Server Performance Evaluation. Proc. of SIGMETRICS (1998)
7. Bhatti N., Friedrich R.: Web Server Support for Tiered Services. IEEE Network (1999) 36–43
8. Casalicchio E., Colajanni M.: A Client-Aware Dispatching Algorithm for Web Clusters Providing Multiple Services. 10th Int'l World Wide Web Conference (2001)
9. Cherkasova L., Phaal P.: Session Based Admission Control: A Mechanism for Improving the Performance of an Overloaded Web Server. Tech Report: HPL-98-119 (1998)
10. Eggert L., Heidemann J.: Application-Level Differentiated Services for Web Servers. World Wide Web Journal (1999) 133–142
11. Parekh S. et al.: Using Control Theory to Achieve Service Level Objectives in Performance Management. Int. Symposium on Integrated Network Management (2001)
12. Glad T., Ljung L.: Reglerteknik: Grundläggande teori (in Swedish). Studentlitteratur (1989)
13. Iyer R., Tewari V., Kant K.: Overload Control Mechansims for Web Servers. Performance and QoS of Next Generation Networks (2000)
14. Lu C. et al.: A Feedback Control Approach for Guaranteeing Relative Delays in Web Servers. Real-Time Technology and Application Symposium (2001)
15. Mindcraft: Webstone, http://www.mindcraft.com
16. Alteon: Alteon Web OS Traffic Control Software. http://www.nortelsnetworks.com/products/01/webos
17. Inktomi: Inktomi Traffic Server C-Class. http://www.inktomi.com/products/cns/products/tscclass_works.html
18. Cisco: Cisco LocalDirector. http://www.cisco.com.
19. Jamjoom H., Reumann J.: QGuard: Protecting Internet Servers from Overload. University of Michigan CSE-TR-427-00 (2000)
20. Bhoj P., Ramanathan S., Singhal S.: Web2K: Bringing QoS to Web Servers. Tech Report: HPL-2000-61 (2000)
21. Challenger J., Dantzig P., Iyengar A.: A Scalable and Highly Available System for Serving Dynamic Data at Frequently Accessed Web Sites. Proc. of ACM/IEEE SC 98 (1998)
22. Manley S., Seltzer M.: Web Facts and Fantasy. USENIX Symposium on Internet Technologies and Systems (1997)
23. Harchol-Balter M., Schroeder B., Agrawal M., Bansal N.: Size-based Scheduling to Improve Web Performance. http://www-2.cs.cmu.edu/~harchol/Papers/papers.html (2002)

24. van de Ven A.: KHTTPd, http://www.fenrus.demon.nl/
25. Aron M., Sanders D., Druschel P., Zwaenepoel W.: Scalable Content-aware Request Distribution in Cluster-based Network Servers. Usenix Annual Technical Conference (2000)
26. Pai V., Aron M., Banga G., Svendsen M., Druschel P., Zwaenepoel W., Nahum E.: Locality-aware Request Distribution in Cluster-based Network Servers. International Conference on Architectural Support for Programming Languages and Operating Systems (1998)
27. Zhang X., Barrientos M., Chen J., Seltzer M.: HACC: An Architecture for Cluster-Based Web Servers. Third Usenix Windows NT Symposium (1999)
28. Aron M., Druschel P., Zwaenepoel W.: Cluster Reserves: a Mechanism for Resource Management in Cluster-based Network Servers. Proc. of ACM SIGMETRICS (2000)
29. Cardellini V., Calajanni M., Yu P.: Dynamic Load Balancing on Web-server Systems. IEEE Internet Computing (1999) 28–39
30. Wang Z.: Cachemesh: A Distributed Cache System for the World Wide Web. 2nd NLANR Web Caching Workshop (1997)
31. Voigt T., Tewari R., Freimuth D., Mehra A.: Kernel Mechanisms for Service Differentiation in Overloaded Web Servers. Usenix Annual Technical Conference (2001)
32. Voigt T., Gunningberg P.: Kernel-based Control of Persistent Web Server Connections, ACM Performance Evaluation Review (2001) 20–25

Dynamic Right-Sizing: An Automated, Lightweight, and Scalable Technique for Enhancing Grid Performance

Wu-chun Feng, Mike Fisk, Mark Gardner, and Eric Weigle

Research & Development in Advanced Network Technology (RADIANT),
Los Alamos National Laboratory, P.O. Box 1663, Los Alamos, NM 87545,
{feng,mfisk,ehw,mkg}@lanl.gov
http://www.lanl.gov/radiant

Abstract. With the advent of computational grids, networking performance over the wide-area network (WAN) has become a critical component in the grid infrastructure. Unfortunately, many high-performance grid applications only use a small fraction of their available bandwidth because operating systems and their associated protocol stacks are still tuned for yesterday's WAN speeds. As a result, network gurus undertake the tedious process of manually tuning system buffers to allow TCP flow control to scale to today's WAN grid environments. And although recent research has shown how to set the size of these system buffers automatically at connection set-up, the buffer sizes are only appropriate at the beginning of the connection's lifetime. To address these problems, we describe an automated and lightweight technique called dynamic right-sizing that can improve throughput by as much as an order of magnitude while still abiding by TCP semantics.

1 Introduction

TCP has entrenched itself as the ubiquitous transport protocol for the Internet as well as emerging infrastructures such as computational grids [1,2], data grids [3,4], and access grids [5]. However, parallel and distributed applications running stock TCP implementations perform abysmally over networks with large bandwidth-delay products. Such large bandwidth-delay product (BDP) networks are typical in grid-computing networks as well as satellite networks [6,7,8].

As noted in [6,7,8,9], adaptation bottlenecks are the primary reason for this abysmal performance, in particular, flow-control adaptation and congestion-control adaptation. In order to address the former problem,[1] grid and network researchers continue to manually tune buffer sizes to keep the network pipe full [7,10,11], and thus achieve acceptable wide-area network (WAN) performance in support of grid computing. However, this tuning process can be quite difficult, particularly for users and developers who are not network experts, because it involves calculating the bandwidth of the bottleneck link and the round-trip

---

[1] The latter problem is beyond the scope of this paper.

G. Carle and M. Zitterbart (Eds.): PfHSN 2002, LNCS 2334, pp. 69–83, 2002.

time (RTT) for a given connection. That is, the optimal TCP buffer size is equal to the product of the bandwidth of the bottleneck link and the RTT, i.e., the bandwidth-delay product of the connection.

Currently, in order to tune the buffer sizes appropriately, the grid community uses diagnostic tools to determine the RTT and the bandwidth of the bottleneck link at any given time. Such tools include *iperf* [12], *nettimer* [13], *netspec* [14], *nettest* [15], *pchar* [16], and *pipechar* [17]. However, none of these tools include a client API, and all of the tools require a certain level of network expertise to install and use.

To simplify the above tuning process, several services that provide clients with the correct tuning parameters for a given connection have been proposed, e.g., AutoNcFTP [18], Enable [19], Web100 [20], to eliminate what has been called the *wizard gap* [21].[2] Although these services provide good first approximations and can improve overall throughput by two to five times over a stock TCP implementation, they only measure the bandwidth and delay at connection set-up time, thus making the implicit assumption that the bandwidth and RTT of a given connection will not change significantly over the course of the connection. In Section 2, we demonstrate that this assumption is tenuous at best. In addition, these services "pollute" the network with extraneous probing packets.

A more dynamic approach to optimizing communication in a grid involves automatically tuning buffers over the lifetime of the connection, not just at connection set-up. At present, there exist two kernel-level implementations: auto-tuning [22] and dynamic right-sizing (DRS) [23,24]. Auto-tuning implements sender-based, flow-control adaptation while DRS implements receiver-based, flow-control adaptation.[3] Live WAN tests show that DRS in the kernel can achieve a 30-fold increase in throughput when the network is uncongested, although speed-ups of 7-8 times are more typical. (And when the network is heavily congested, DRS throttles back and performs no better than the default TCP.) However, achieving such speed-ups requires that our kernel patch for DRS be installed in the operating systems of every pair of communicating hosts in a grid.[4]

The installation of our DRS kernel patch requires knowledge about adding modules to the kernel and *root* privilege to install the patch. Thus, the DRS functionality is generally not accessible to the typical end user (or developer). However, in the longer term, we anticipate that this patch will be incorporated into the kernel core so that its installation and operation are transparent to the end user. In the meantime, end users still demand the better performance of DRS but with the pseudo-transparency of Enable and AutoNcFTP. Thus, we propose a more portable implementation of DRS in *user space* that is transpar-

---

[2] The *wizard gap* is the difference between the network performance that a network "wizard" can achieve by appropriately tuning buffer sizes and the performance of an untuned application.

[3] The Web100 project recently incorporated DRS into their software distribution to enable the dynamic sizing of flow-control windows over the lifetime of a connection [25].

[4] Once installed, not only do grids benefit, but every TCP-based application benefits, e.g., *ftp*, multimedia streaming, WWW.

ent to the end user. Specifically, we integrate our DRS technique into *ftp* (drs-FTP). The differences between our drsFTP and AutoNcFTP are two-fold. First, AutoNcFTP relies on NcFTP (http://www.ncftp.com/) whereas drsFTP uses a de-facto standard *ftp* daemon from Washington University (http://www.wu-ftpd.org/). Second, the buffers in AutoNcFTP are only tuned at connection set-up while drsFTP buffers are dynamically tuned over the lifetime of the connection, thus resulting in better adaptation and better overall performance.

The remainder of the paper is organized as follows. Section 2 demonstrates why dynamic, flow-control adaptation is needed over the lifetime of the connection rather than at connection set-up only. Sections 3 and 4 describe the DRS technique and its implementation in kernel space and in user space, respectively. Then, in Section 5, we present our experimental results, followed by concluding remarks in Section 6.

2   Background

TCP relies on two mechanisms to set its transmission rate: flow control and congestion control. Flow control ensures that the sender does not overrun the receiver's available buffer space (i.e., a sender can send no more data than the size of the receiver's last advertised flow-control window) while congestion control ensures that the sender does not unfairly overrun the network's available bandwidth. TCP implements these mechanisms via a flow-control window ($fwnd$) that is advertised by the receiver to the sender and a congestion-control window ($cwnd$) that is adapted based on inferring the state of the network.

Specifically, TCP calculates an effective window ($ewnd$), where $ewnd \equiv min(fwnd,cwnd)$, and then sends data at a rate of $ewnd/RTT$, where RTT is the round-trip time of the connection. Currently, $cwnd$ varies dynamically as the network state changes; however, $fwnd$ has always been static despite the fact that today's receivers are not nearly as buffer-constrained as they were twenty years ago. Ideally, $fwnd$ should vary with the bandwidth-delay product (BDP) of the network, thus providing the motivation for DRS.

Historically, a static $fwnd$ sufficed for all communication because the BDP of networks was small. Hence, setting $fwnd$ to small values produced acceptable performance while wasting little memory. Today, most operating systems set $fwnd \approx 64$ KB – the largest window available without scaling [26]. Yet BDPs range between a few bytes (56 Kbps × 5 ms → 36 bytes) and a few megabtyes (622 Mbps × 100 ms → 7.8 MB). For the former case, the system wastes over 99% of its allocated memory (i.e., 36 B / 64 KB = 0.05%). In the latter case, the system potentially wastes up to 99% of the network bandwidth (i.e., 64 KB / 7.8 MB = 0.80%).

Over the lifetime of a connection, bandwidth and delay change (due to transitory queueing and congestion) implying that the BDP also changes. Figures 1, 2, and 3 support this claim.[5] Figure 1 presents the bottleneck bandwidth between Los Alamos and New York at 20-second intervals. The bottleneck bandwidth

---

[5] To generate these figures, we used *nettimer* to measure bandwidth and RTT delay.

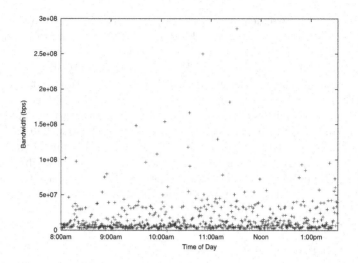

Fig. 1. Bottleneck Bandwidth at 20-Second Intervals

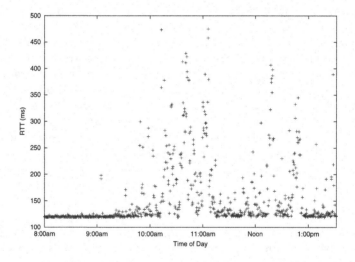

Fig. 2. Round-Trip Time at 20-Second Intervals

averages 17.2 Mbps with a low and a high of 0.026 Mbps and 28.5 Mbps, respectively. The standard deviation and half-width of the 95% confidence interval are 26.3 Mbps and 1.8 Mbps. Figure 2 shows the RTT, again between Los Alamos and New York, at 20-second intervals. The RTT delay also varies over a wide range of [119, 475] ms with an average delay of 157 ms. Combining Figures 1 and 2 results in Figure 3, which shows that the BDP of a given connection can vary by as much as 61 Mb.

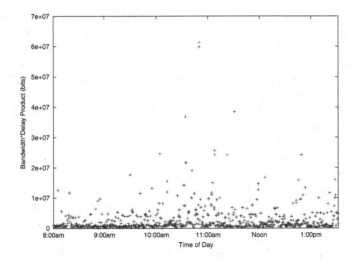

Fig. 3. Bandwidth-Delay Product at 20-Second Intervals

Based on the above results, the BDP over the lifetime of a connection is continually changing. Therefore, a fixed value for $fwnd$ is not ideal; selecting a fixed value forces an implicit decision between (1) under-allocating memory and under-utilizing the network or (2) over-allocating memory and wasting system resources. Clearly, the grid community needs a solution that dynamically and transparently adapts $fwnd$ to achieve good performance without wasting network or memory resources.

3   Dynamic Right-Sizing (DRS) in the Kernel

Dynamic right-sizing (DRS) lets the receiver estimate the sender's $cwnd$ and then use that estimate to dynamically change the size of the receiver's window advertisements $fwnd$ (as memory resources will allow on the receiver side). These updates can then be used to keep pace with the growth in the sender's congestion window. As a result, the throughput between end hosts, e.g., as in a grid, will only be constrained by the available bandwidth of the network rather than some arbitrarily set constant value on the receiver that is advertised to the sender.

Initially, at connection set-up, the sender's $cwnd$ is smaller than the receiver's advertised window $fwnd$. To ensure that a given connection is not flow-control constrained, the receiver must continue to advertise a $fwnd$ that is larger than the sender's $cwnd$ before the receiver's next adjustment.

The instantaneous throughput seen by a receiver may be larger than the available end-to-end bandwidth. For instance, data may travel across a slow link only to be queued up on a downstream router and then sent to the receiver in one or more fast bursts. The maximum size of such a burst is bounded by the size of the sender's $cwnd$ and the window advertised by the receiver. Because the sender can send no more than one *ewnd* window's worth of data between acknowledgements, a burst that is shorter than a RTT can contain at most one *ewnd*'s worth of data. Thus, for any period of time that is shorter than a RTT, the amount of data seen over that period is a *lower bound* on the size of the sender's *cwnd*. But how does such a distributed system calculate its RTT?

In a typical TCP implementation, the RTT is measured by observing the time between when data is sent and an acknowledgement is returned. However, during a bulk-data transfer (e.g., from sender to receiver), the receiver may not be sending any data, and therefore, will not have an accurate RTT estimate. So, how does the receiver infer delay (and bandwidth) when it only has acknowledgements to transmit back and no data to send?

A receiver in a computational grid that is only transmitting acknowledgements can still estimate the RTT by observing the time between when a byte is first acknowledged and the receipt of data that is at least one window beyond the sequence number that was acknowledged. If the sending application does not have any data to transmit, the measured RTT could be much larger than the actual RTT. Thus, this measurement acts as an *upper bound* on the RTT and should be used only when it is the only source of RTT information.

For a more rigorous and mathematical presentation for the lower and upper bounds that are used in our kernel implementation of DRS, please see [23,24].

4   DRS in User Space: drsFTP

Unlike the kernel-space DRS, user-space DRS implementations are specific to a particular application. Here, we integrate DRS into `ftp`, resulting in drsFTP. As with AutoNcFTP and Enable, we focus on (1) adjusting TCP's system buffers over the data channel of `ftp` rather than the control channel and (2) using `ftp`'s stream file-transfer mode. The latter means that a separate data connection is created for every file transferred. Thus, during the lifetime of the transfer, the sender *always* has data to transmit; once the file has been completely sent, the data connection closes. We leverage the above information in our design of drsFTP.

The primary difficulty in developing user-space DRS code lies in the fact that user-space code generally does *not* have direct access to the high-fidelity information available in the TCP stack. Consequently, drsFTP has no knowledge of parameters generated by TCP such as the RTT of a connection or the receiver's advertised window.

4.1   Adjusting the Receiver's Window

Because the receiver is running in user space, it is unable to determine the actual round trip time of TCP packets. However, in developing drsFTP, we do know that the sender always has data to send for the life of the connection. It then follows that the sender will send as much data as possible, limited by its idea of congestion- and flow-control windows. So, the receiver can assume that it is receiving data as quickly as the current windows and network conditions allow.

We use this assumption in the following manner. The drsFTP application maintains an application receive buffer of at least twice the size of the current kernel TCP receive buffer (which can be obtained via the `getsockopt()` function call with the `TCP_RCVBUF` parameter). Every time the application reads from the network, it attempts to read an entire receive buffer's worth of data. If more data than some threshold value is read, the assumption can be made that the flow-control window should be increased.

The threshold value depends on the operating system (OS), in particular, how much of the TCP kernel buffer that the OS reserves as application buffer space and how much is used for the TCP transfer (i.e., how much of the buffer is used for the TCP receive window). The threshold value is always less than the value reported by `TCP_RCVBUF`. For example, we based our tests of drsFTP on the Linux operating system. Linux maintains half of the TCP buffer space as the TCP receive window and half as buffer area for the application. Thus, the threshold value we used was $\frac{1}{2}$ the total reported by `TCP_RCVBUF` size.

In the worst case, the sender's window is doubling with every round trip (i.e., during TCP slow start). Thus, when the determination is made that the receiver window should increase, the new value should be at least double the current value. In addition, the new value should take the threshold value into consideration. Thus, for our drsFTP implementation, we increase the receive window by a factor of four. This factor is applied to both the application buffer and the kernel buffer (via `setsockopt()/TCP_RCVBUF`).

4.2 Adjusting the Sender's Window

In order to take full advantage of dynamically changing buffer sizes, the sender's buffer should adjust in step with the receiver's. This presents a problem in user-space implementations because the sender's user-space code has no way of determining the receiver's advertised TCP window size. However, the `ftp` protocol specification [27] provides a solution to this dilemma. Specifically, `ftp` maintains a control channel, which is a TCP connection completely separate from the actual data transfer. Commands are sent from the client to the server over this control channel, and replies are sent in the reverse direction. Additionally, the `ftp` specification does not prohibit traffic on the control channel during data transfer. Thus, a drsFTP receiver may inform a drsFTP sender changes in buffer size by sending appropriate messages over the `ftp` control channel.

Since `ftp` is a bidirectional data-transfer protocol, the receiver may either be the `ftp` server or client. However, RFC 959 specifies that only `ftp` clients may send commands on the control channel, while `ftp` servers may only send replies to commands. Thus, a new `ftp` command and reply must be added to the `ftp` implementation in order to fully implement drsFTP. Serendipitously, the Internet Draft of the GridFTP protocol extensions to `ftp` defines an `ftp` command "SBUF", which is designed to allow a client to set the server's TCP buffer sizes before data transfer commences. We extend the definition of SBUF to allow this command to be specified during a data transfer, i.e., to allow buffer sizes to be set dynamically. The full definition of the expanded SBUF command appears below:

Syntax:

```
sbuf = SBUF <SP> <buffer-size>
buffer-size ::= <number>
```

This command informs the server-PI to set the TCP buffer size to the value specified (in bytes). SBUF may be issued at any time, including before or during active data transfer. If specified during data transfer, it affects the data transfer that started most recently.

Response Codes:

If the server-PI is able to set the buffer size to the requested buffer size, a 200 response code may be returned. No response code is necessary if specified during a data transfer, but a response is required if specified outside of the data transfer.

In addition, we propose a new reply code to allow the server-as-receiver to notify the client of changes in the receiver window.

```
126 Buffer Size (xxx)
xxx ::= buffer size in bytes
```

The 126 Reply may occur at any point when the server-PI is sending data to the user-PI (or a server-PI running concurrently with the user-PI). As with the SBUF command during data transfer, this reply is informational and need not be acted upon or responded to in any manner.

This reply code is consistent with RFC 959 and does not interfere with any ftp extension or proposed extension.

4.3   TCP Window Scaling

Because the window-scaling factor in TCP is established at connection set-up time, an appropriate scale must be set before a new data connection is opened. Most operating systems allow TCP_RCVBUF and TCP_SNDBUF to be set on a socket before a connection attempt is made and then use the requested buffer size to establish the TCP window scaling.

5   Experiments

In this section, we present results for both the kernel- and user-space implementations of DRS. In particular, we will show that the throughput for both the kernel- and user-space implementations improves upon the default configuration by 600% and 300%, respectively. The kernel implementation performs better because it has access to fine-granularity information in the kernel and has two fewer copies to perform than drsFTP.

5.1   Experimental Setup

Our experimental apparatus, shown in Figure 4, consists of three identical machines connected via Fast Ethernet. Each machine contains a dual-CPU 400-MHz Pentium II with 128-MB of RAM and two network-interface cards (NICs). One machine acts as a WAN emulator with a 100-ms round-trip time (RTT) delay; each of its NICs is connected to one of the other machines via crossover cables (i.e., no switch).

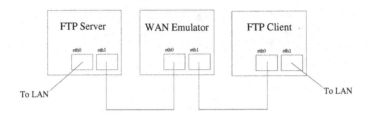

Fig. 4. Experimental Setup

5.2   Kernel-Space DRS

In the kernel implementation of DRS, the receiver estimates the size of the sender's congestion window so it can advertise an appropriate flow-control window to the sender. Our experiments show that the DRS algorithm approximates the actual size quite well. Further, we show that by using this estimate to size the window advertisements, DRS keeps the connection congestion-control limited rather than (receiver) flow-control limited.

Performance. As expected, using larger flow-control windows significantly enhances WAN throughput versus using the default window sizes of TCP. Figure 5 shows the results of 50 transfers of 64 MB of data with `ttcp`, 25 transfers using the default window size of 32 KB for both the sender and receiver and 25 transfers using DRS. Transfers with the default window sizes took a median time of 240 seconds to complete while the DRS transfers only took 34 seconds (or roughly *seven* times faster).

Figures 6 and 7 trace the window size and flight size of the default-sized TCP and the DRS TCP. (The flight size refers to the amount of sent but unacknowledged data in the sender's buffer. This flight size, in turn, is bounded by the window advertised by the receiver.) For the traditionally static, default,

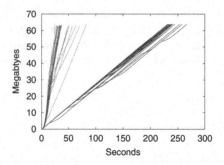

Fig. 5. Progress of Data Transfers

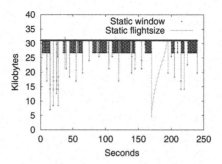

Fig. 6. Default Window Size: Flight & Window Sizes

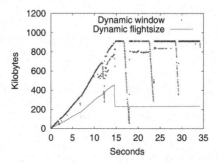

Fig. 7. Dynamic Right-Sizing: Flight & Window Sizes

flow-control window as shown in Figure 6; the congestion window quickly grows and becomes rate-limited by the receiver's small 32-KB advertisement for the flow-control window. On the other hand, DRS allows the receiver to advertise a window size that is roughly twice the largest flight size seen to date (in case, the connection is in slow start). Thus, the flight size is only constrained by the conditions in the network, i.e., congestion window. Slow start continues for much longer and stops only when packet loss occurs. At this point, the congestion window stabilizes on a flight size that is roughly seven times higher than the constrained flight size of the static case. And not coincidentially, this seven-fold increase in the average flight size translates into the same seven-fold increase in throughput shown in Figure 5.

In additional tests, we occasionally observe increased queueing delay caused by the congestion window growing larger than the available bandwidth. This causes the retransmit timer to expire and reset the congestion window to one even though the original transmission of the packet was acknowledged shortly thereafter.

Low-Bandwidth Connections. Figures 8 and 9 trace the window size and flight size of default-sized TCP and DRS TCP over a 56K modem. Because DRS provides the sender with indirect feedback about the achieved throughput rate, DRS actually causes a TCP Reno sender to induce *less* congestion and fewer retransmissions over bandwidth-limited connections. Although the overall throughput measurements for both cases are virtually identical, the static (default) window generally has more data in flight as evidenced by the roughly 20% increase in the number of re-transmissions shown in Figure 10. This ad-

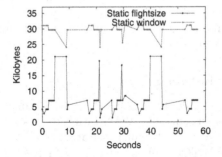

Fig. 8. Default Window Size: Low-Bandwidth Links (56K Modem)

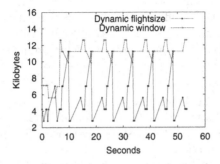

Fig. 9. Dynamic Right-Sizing: Low-Bandwidth Links (56K Modem)

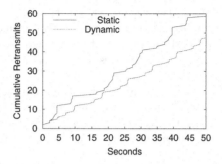

Fig. 10. Retransmissions in Low-Bandwidth Links

ditional data in flight is simply dropped because the link cannot support that throughput.

Discussion. The Linux 2.4.x kernel contains complementary features to DRS that are designed to reduce memory usage on busy web servers that are transmitting data on large numbers of network-bound TCP connections. Under normal circumstances, the Linux 2.4 kernel restricts each connection's send buffers to be just large enough to fill the current congestion window. When the total memory used exceeds a threshold, the memory used by each connection is further constrained. Thus, while Linux 2.4 precisely bounds send buffers, DRS precisely bounds receiver-side send buffers.

5.3   drsFTP: DRS in User Space

For each version of `ftp` (drsFTP, stock FTP, statically-tuned FTP), we transfer 100-MB files over the same emulated WAN with a 100-ms RTT delay. As a baseline, we use stock FTP with TCP buffers set at 64 KB. Most modern operating systems set their default TCP buffers to 64 KB, 32 KB, or even less. Therefore, this number represents the high-end of OS-default TCP buffer sizes. We then test drsFTP, allowing the buffer size to vary in response to network conditions while starting at 64 KB as in stock FTP. Lastly, we benchmark a statically-tuned FTP, one that tunes TCP buffers once at connection set-up time. To test the extremes of this problem, we test the statically-tuned FTP with two different values – one set to the minimum bandwidth-delay product (BDP) and one to the maximum.

Performance. Figure 11 shows the average time to transfer a 100-MB file along with the range of the half-width of the 95% confidence interval centered around the average. Both the drsFTP and statically-tuned FTP produce a four-fold improvement over stock FTP.

When the minimum values are used for the statically-tuned FTP to set $fwnd$, the actual BDP fluctuates significantly below and above the set value over the

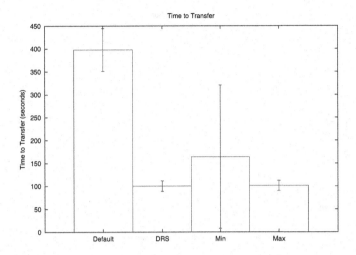

Fig. 11. Transfer Time (Smaller is better)

course of the file transfer. This results in a very large variability in transfer times, as shown in Figure 11. However, the memory usage stays relatively low when compared to the statically-tuned FTP where maximum values are used, as shown in Figure 12.

When the maximum values are used, the transfer time is competitive with drsFTP but at the expense of using a tremendous amount of memory. That is, the buffer sizes are set large enough that the statically set value of $fwnd$ is never exceeded.

As shown in Figures 11 and 12, drsFTP simultaneously achieves fast transfer times (comparable to statically-tuned FTP with maximum values and four times faster than stock FTP) and relatively low-memory usage (comparable to the statically-tuned FTP with minimum values).

6   Conclusion

In this paper, we presented dynamic right-sizing (DRS), an automated, light-weight, and scalable technique for enhancing grid performance. Over a typical WAN configuration, the kernel-space DRS achieves a seven-fold increase in throughput versus stock TCP; and the user-space DRS, i.e., drsFTP, achieves a four-fold increase in throughput versus stock TCP/FTP.

Currently, the biggest drawback of drsFTP is its double-buffered implementation, i.e., buffered in the kernel and then in user space. Clearly, this implementation affects the throughput performance of drsFTP as drsFTP achieves only 57% of the performance of the kernel-space DRS. However, this implementation of drsFTP was simply a proof-of-concept to see if it would provide any benefit over a statically-tuned FTP. Now that we have established that there is sub-

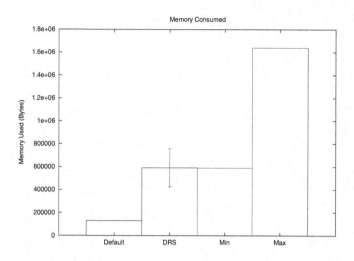

Fig. 12. Memory Usage (Smaller is better.)

stantial benefit in implementing DRS in user space, we are currently working on a new version of drsFTP that abides more closely to our kernel implementation of DRS (albeit at a coarser time granularity) and eliminates the extra copying done in the current implementation of drsFTP.

Lastly, we are working with Globus middleware researchers to integrate our upcoming higher-fidelity version of drsFTP with GridFTP. Currently, GridFTP uses parallel streams to achieve high bandwidth.

References

1. Foster, I. and Kesselman, C.: The Grid: Blueprint for a New Computing Infrastructure. Morgan-Kaufmann Publishers. San Francisco, California (1998).
2. Foster, I., Kesselman, C., Tuecke, S.: The Anatomy of the Grid: Enabling Scalable Virtual Organizations. International Journal of Supercomputer Applications (2001).
3. Chervenak, A., Foster, I., Kesselman, C., Salisbury, C., Tuecke, S.: The Data Grid: Towards an Architecture for the Distributed Management and Analysis of Large Scientific Datasets. International Journal of Supercomputer Applications (2001).
4. Argonne National Laboratory, California Institute of Technology, Lawrence Berkeley National Laboratory, Stanford Linear Accelerator Center, Jefferson Laboratory, University of Wisconsin, Brookhaven National Laboratory, Fermi National Laboratory, and San Diego Supercomputing Center: The Particle Physics Data Grid. http://www.cacr.caltech.edu/ppdg/.
5. Childers, L., Disz, T., Olson, R. Papka, M., Stevens, R., Udeshi, T.: Access Grid: Immersive Group-to-Group Collaborative Visualization. Proceedings of the 4th International Immersive Projection Workshop (2000).
6. Partridge, C., Shepard, T.: TCP/IP Performance over Satellite Links. IEEE Network. **11** (1997) 44–49.

7. Allman, M., Glover, D., Sanchez, L.: Enhancing TCP Over Satellite Channels Using Standard Mechanisms. IETF RFC 2488 (1999).
8. Allman, M. et al.: Ongoing TCP Research Related to Satellites. IETF RFC 2760 (2000).
9. Feng, W., Tinnakornsrisuphap, P.: The Failure of TCP in High-Performance Computational Grids. Proceedings of SC 2000: High-Performance Networking and Computing Conference (2000).
10. Pittsburgh Supercomputing Center. Enabling High-Performance Data Transfers on Hosts. http://www.psc.edu/networking/perf_tune.html.
11. Tierney, B. TCP Tuning Guide for Distributed Applications on Wide-Area Networks. USENIX & SAGE Login. http://www-didc.lbl.gov/tcp-wan.html (2001).
12. Tirumala, A. and Ferguson, J.: IPERF Version 1.2. http://dast.nlanr.net/Projects/Iperf/index.html (2001).
13. Lai, K., Baker, M.: Nettimer: A Tool for Measuring Bottleneck Link Bandwidth. Proceedings of the USENIX Symposium on Internet Technologies and Systems (2001).
14. University of Kansas, Information & Telecommunication Technology Center: NetSpec: A Tool for Network Experimentation and Measurement. http://www.ittc.ukans.edu/netspec.
15. Lawrence Berkeley National Laboratory: Nettest: Secure Network Testing and Monitoring. http://www-itg.lbl.gov/nettest.
16. Mah, B.: pchar: A Tool for Measuring Internet Path Characteristics. http://www.employees.org/ bmah/Software/pchar.
17. Jin., G., Yang, G., Crowley, B., Agrawal, D.: Network Characterization Service. Proceedings of the IEEE Symposium on High-Performance Distributed Computing (2001).
18. Liu, J., Ferguson, J.: Automatic TCP Socket Buffer Tuning. Proceedings of SC 2000: High-Performance Networking and Computing Conference (2000).
19. Tierney, B., Gunter, D., Lee, J., Stoufer, M.: Enabling Network-Aware Applications. Proceedings of the IEEE International Symposium on High-Performance Distributed Computing (2001).
20. National Center for Atmospheric Research, Pittsburgh Supercomputing Center, and National Center for Supercomputing Applications. The Web100 Project. http://www.web100.org.
21. Mathis, M.: Pushing Up Performance for Everyone. http://ncne.nlanr. net/training/techs/1999/991205/Talks/ mathis_991205_Pushing_Up_Performance/ (1999).
22. Semke, J., Mahdavi, J., Mathis, M. Automatic TCP Buffer Tuning. Proceedings of ACM SIGCOMM (1998).
23. Fisk, M., Feng, W.: Dynamic Adjustment of TCP Window Sizes. Los Alamos National Laboratory Unclassified Report, LA-UR 00-3221 (2000).
24. Fisk, M., Feng, W.: Dynamic Right-Sizing: TCP Flow-Control Adaptation (poster). Proceedings of SC 2001: High-Performance Networking and Computing Conference (2001).
25. Dunigan, T., Fowler, F. Personal Communication with Web100 Project (2002).
26. Jacobson, V., Braden, R., Borman, D.: TCP Extensions for High Performance. IETF RFC 1323 (1992).
27. Postel, J., Reynolds, J. File Transfer Protocol (FTP). IETF RFC 959 (1985).

The "Last-Copy" Approach for Distributed Cache Pruning in a Cluster of HTTP Proxies*

Reuven Cohen and Itai Dabran

Technion, Haifa 32000, Israel

Abstract. Web caching has been recognized as an important way to address three main problems in the Internet: network congestion, transmission cost and availability of web servers. As traffic increases, cache clustering becomes a natural way to increase scalability. This paper proposes an efficient scheme for increasing the cache hit-ratio in a loosely-coupled cluster. In such a cluster, each proxy is able to serve every request independently of the other proxies. In order to increase the performance, the proxies may share cacheable content using some inter-cache communication protocol. The main contribution of the proposed scheme is an algorithm that increases the performance (hit-ratio) of any cache-pruning algorithm in such a cluster.

1   Introduction

WWW's main problems fall into 3 categories: (a) Internet congestion delays; (b) transmission cost, and (c) availability of web servers. Web caching has been recognized as an important way to address these problems. A web proxy cache sits between the Web servers and clients, and stores frequently accessed web objects. The proxy receives requests from the clients and when possible serves these requests using the stored objects. This concept helps the end user, the service provider and the content provider by reducing the server load, alleviating network congestion, reducing bandwidth consumption and reducing network latency.

Cache clustering is a natural way to scale as traffic increases. The idea is to replace a single cache, when it is no longer able to handle all the traffic, with a cluster of caches. However, cache clustering introduces a new "request-routing problem": which specific cache in the cluster should address each request received from the users. There are several approaches to address this problem, the most popular two are as follows:

1. The URL-space is partitioned, usually through a hash function, across the cluster. Each cache in the cluster is responsible only for the web items whose URLs are hashed to it. The browser, or a proxy in another cluster, hashes the URL, and directs the request message to a specific proxy that is most

---

* This work was supported by a grant from the Israeli Ministry of Science, Culture and Sport (MOS)

G. Carle and M. Zitterbart (Eds.): PfHSN 2002, LNCS 2334, pp. 84–99, 2002.

likely to have the required item. The relationship between the caches in such architecture is referred to as *tightly-coupled* [3]. The Cache Array Routing Protocol (CARP) supports this approach.

2. The cluster is arranged in a *loosely-coupled* manner [3]. Each proxy in the cluster is able to serve every request, independently of the other proxies. In order to increase the performance, the proxies may share cacheable content using some inter-cache communication protocol. The most popular protocol for loosely-coupled inter-proxy communication are ICP (Internet Cache Protocol) [14] and HTCP (Hyper Text Caching Protocol) [11].

The main advantages of the tightly-coupled approach over the loosely-coupled approach is that there is no need for inter-cache communication protocol, since a request for a web item is "automatically" routed to the correct proxy. The main advantages of the loosely-coupled approach are attributed to the fact that each proxy server can serve each web item, and therefore (a) this approach can guarantee a much better load balancing; (b) scaling, by adding additional cache proxies, is easy; and (c) this approach is less sensitive to proxy failures.

This paper deals with a loosely-coupled cluster of cache proxies. There are several approaches for "routing" incoming requests. Most of these approaches employ a special logic, referred to as a "director" (or a "dispatcher"), that sits in the entrance of the cluster. The director receives each request, and determines the specific proxy to which this request should be forwarded. Generally, two types of directors are used: a Layer-5 (Application Layer) director and a Layer-4 (Transport Layer) director. The former makes request-routing decisions based on the Application Layer data, like the required URL, cookies, etc., whereas the latter makes these decisions with no awareness to the Application Layer data. The main approach for implementing a Layer-5 director is using another proxy server. Such a proxy terminates the TCP connection with the client, views and analyzes the client request, determines the best proxy machine within the cluster that can serve the request, and forwards the request to this proxy over another TCP connection. In contrast, a Layer-4 director determines the target machine upon receiving the client TCP SYN segment, and therefore without reading the request data. It determines the most available cache proxy, and assigns the new connection to this machine. This usually involves changing the destination IP address of the connection's packets to the internal IP address of that machine through a process known as NAT (Network Address Translation) [9].

Both director types have advantages and disadvantages. The main advantage of a Layer-4 director is that it is much less loaded: unlike a Layer-5 director, that needs to participate in two TCP connections in order to serve each request, the Layer-4 director needs to perform only NAT and IP routing. The main advantage of a Layer-5 director is that it may make smarter request-routing decisions, based on the content of the requests which is unknown to a Layer-4 director. This may help in achieving better load-balancing, and in increasing the cache hit-ratio probability.

A Layer-4 director is the most common approach to route requests to the proxies of a loosely-coupled cluster with no regard to the content of the requests.

However, other approaches exist as well. For instance, another common approach is to use a round-robin DNS in order to send the browser of each user a different IP address for the same cache proxy name.

This paper proposes an efficient scheme for increasing the cache hit-ratio in a loosely-coupled cluster, assuming that the request-routing decision is made by a Layer-4 director, or by any other mechanism that is *unaware of the content of the requests*. Under the considered model a received request is routed to one of the cluster proxies, regardless of the content of the request, either randomly or according to some scheduling policy like round-robin. In a system that employs a Layer-5 director, the director may maintain information regarding the proxy that holds the requested object, in which case the problem is different. The main part of the proposed scheme is an algorithm that increases the performance of any cache pruning algorithm, when the latter is implemented in a cluster of cache servers. A cache pruning algorithm is periodically invoked in order to remove web items from a cache until the cache free space reaches some pre-determined threshold. Cache pruning logic for traditional CPU or I/O subsystems usually makes a decision based on one of the following two parameters:

(1) LRU (Least Recently Used): the last time each item was accessed is considered; the least recently accessed items are assigned a higher removal priority.
(2) LFU (Least Frequently Used): the number of references to each item during the last period of time is considered; the least referenced items are assigned a higher removal priority.

Due to the heterogeneity of the web, the delay variance associated with retrieving a web item from an original server is very large. This gives rise to pruning algorithms that took into account another important parameter, namely

(3) The time it takes to retrieve each item from its original server [10][17]: a web item that can be more rapidly retrieved is assigned a higher removal priority.

However, in a loosely-coupled cluster of proxies, the pruning algorithm executed by each individual proxy should take into consideration another important parameter:

(4) Whether the web item exists or not exists in another proxy within the same cluster. Since we consider a LAN-based, rather than a WAN-based cluster, the cost (in terms of delay and bandwidth) for bringing a web item from another proxy in the same cluster is significantly smaller than the cost for bringing it from the original server. Therefore, items that exist in another proxy within the same cluster should be assigned a higher removal priority.

Designing a pruning algorithm for a loosely-coupled cluster of proxies that takes into consideration all the parameters listed above is difficult for three main reasons:

– The problem of finding the best item to be removed based on two or more of the parameters discussed above is NP-complete [10].

- Estimating the time it takes for bringing a web item from the origin server is a difficult problem, mainly because the load on the servers and on the network may vary rapidly.
- In a loosely-coupled cluster with a Layer-4 director, it is difficult for one proxy to know which web items the other proxies in the cluster hold. A scheme that addresses this problem is presented in [4] and [7]. In this scheme each proxy maintains a compact summary of the web items each other proxy stores. A proxy uses this information in order to identify a local proxy that stores an item for which a request is received. However this approach requires a periodical summary update of each proxy's content, and additional storage capacity at each of the proxies in order to store the cache summaries.

In this paper we address these difficulties using a concept of "Last-Copy" policy, which is enforced by many university libraries. According to this concept, one copy of each important text-book is stamped as "Last-Copy". This copy can be usually used only in the library, in order to increase its availability to someone who really needs it.

We adopt this idea to our scheme in the following way. The cluster may have multiple copies of each web item. However, exactly one copy is marked as "Last-Copy". When the pruning algorithm is executed by each cache proxy, the proxy distinguishes between web items that are marked as "Last-Copy" and web items that are not marked as "Last-Copy". The items from the first group are assigned a much lower pruning priority than the items from the second group, and they are therefore more likely to remain in the cache after the pruning phase terminates. Therefore, regardless of the pruning algorithm employed in the cluster, the concept guarantees with high probability that at least one copy of each item remains in the cluster.

The paper also proposes an ICP-like cache sharing scheme. This scheme ensures that indeed one, and only one, copy of each item is marked as "Last-Copy". The scheme includes a protocol that allows a cluster proxy to search for a web item that is not available in its local cache, and to get a copy of this item from the proxy that holds the "Last-Copy" of this item.

The rest of the paper is organized as follows. In Section 2 we discuss cache pruning algorithms for a single proxy. In Section 3 we discuss this problem in the context of a cluster of proxies. We present our "Last-Copy" scheme and show how this scheme can be integrated into an algorithm for a single proxy in order to work more efficiently in an environment with a cluster of proxies. In Section 4 we present an ICP-like intra-cluster cache sharing protocol, based on the "Last-Copy" concept. In Section 5 we present simulation results that demonstrate the advantages of the proposed scheme, and in Section 6 we conclude the paper.

2   The Cache Pruning Problem

Caching gain can be measured as an Object-Hit-Ratio or as Byte-Hit-Ratio. Object-Hit-Ratio is the ratio of the number of requests served from the cache to the total number of requests. Byte-Hit-Ratio is the ratio of the number of bytes served from the cache to the total number of served bytes. A higher Object-Hit-Ratio indicates higher success in retrieving web items from the cache, whereas a higher Byte-Hit-Ratio indicates a better response to requests of relatively large items. A cache pruning policy that seeks to optimize the Object-Hit-Ratio usually removes large items from the cache, whereas a policy that seeks to optimize the Byte-Hit-Ratio usually does not take into consideration the size of the items. Traditional cache algorithms for CPU or I/O subsystems are based on the concept of "locality of reference", and therefore usually employ an LRU (Least Recently Used) based pruning algorithm. Since CPU or I/O items usually have a fixed size and an equal retrieving time, for such subsystems there is no difference between the Object-Hit-Ratio and the Byte-Hit-Ratio. In contrast, web items have different size and variable retrieving time. Therefore, as explained in Section 1, web cache pruning algorithms take into account not only LRU or LFU considerations, but also the time it takes to obtain the web item from its original web server. Moreover, in order to increase the Object-Hit-Ratio, sometimes on the expense of decreasing the Byte-Hit-Ratio, a pruning algorithm take into account also the size of the page and prune from the cache one big item rather than multiple small items.

In [17] it was found that the performance of a cache pruning algorithm based on the retrieval time only, namely removing the pages whose retrieval time is minimum, is worst than the results achieved by traditional LRU and LFU policies. According to [17], an optimal policy should combine the following 4 parameters: (1) the network latency between the client and the server, which mainly affects the time to open the TCP connection; (2) the bandwidth to the original server; (3) the size of the item; and (4) the popularity of the item, as reflected by LFU. The Hybrid algorithm proposed in [17] keeps a per-origin-server statistics of the estimated connection setup time with this server, and the estimated bandwidth on the path to this server. The cache proxy needs to measure these parameters and to update them each time it accesses the server.

In [15] it is proposed to remove items according to their size only. When the proxy is full, the largest web item is removed. As expected, in [15] and [16] it is argued that this policy indeed maximizes the Object-Hit-Ratio, but has a negative effect on the Byte-Hit-Ratio.

Another work [6] presents the LRV (Lowest Relative Value) policy. This policy combines the parameters of LRU, LFU and object size. According to this algorithm, the removal priority of an item is a function of the time since the last access to this item, the number of accesses to this item during the last Δ seconds, and the size of the item.

A LWU (Least Weighted Usage) removal policy, is presented in [8]. LWU combines only the parameters of LRU and LFU. In experiments using two web server logs, it was found that LWU performs better than LRU and LFU. LRU-MIN [1], is a variant of LRU that tries to minimize the number of items to be replaced. Another LRU variation is LRU-THOLD [1]. This algorithm prevents large web items from entering the cache at all. It was found to perform well for small caches.

3   The "Last-Copy" Concept

In Section 2 we have discussed the most popular cache pruning algorithms for a single proxy. These algorithms are referred in what follows to as *single-proxy (or single-cache) pruning algorithms*. In this section we address the problem of cache pruning in a cluster of cache proxies. Very often, in such a system this issue is addressed by implementing a single proxy pruning algorithm in each of the cluster proxies independently of the other proxies. However, we show that better performance is achieved when the pruning algorithm of each cache takes into account not only the status of the local cache, but also the status of the caches in the other proxies of the cluster. Certainly, in a tightly-coupled system, a global pruning algorithm that takes into account the exact status of each cache can be developed. However, as already noted this paper deals with a loosely-coupled system, where each proxy works independently of the other proxies in the cluster. Moreover, we consider a Layer-4 director. Recall that such a director is unaware of the content of each request, and is therefore unable to synchronize the caches of the various cluster proxies.

In order to take into consideration the content of the other cache proxies in the cluster, we employ the concept of "Last-Copy". As already indicated, this idea is borrowed from a common policy in university libraries, where one copy of each important textbook is stamped with the "Last-Copy" label. The library imposes a special policy on this copy, in order to improve its availability compared to the other copies of the same book. Our "Last-Copy"-based scheme can be combined with almost every single-proxy pruning algorithm, and in particular with the most common algorithms presented in Section 2. The main idea is as follows. Assume a mechanism that guarantees that exactly one copy of every web item in the cluster is marked as Last-Copy. When the proxy runs its single-proxy pruning algorithm, it assigns a lower removal priority to each local item marked as "Last-Copy". The purpose is to guarantee that even if an item is pruned from the cache of the other proxies in the cluster, its "Last-Copy" remains.

We now show how this concept can be integrated into the most common single-proxy pruning algorithms that take into consideration the time since the last access to an item (LRU) and the number of accesses to an item (LFU). Each proxy maintains two groups of items, a group of "Last-Copy" items and a group of "Not-Last-Copy" items. In the simple version of the scheme, the proxy only removes items from the "Not-Last-Copy" items list, based on the rules of the single-proxy pruning algorithm. Consequently this approach assigns the "Last-Copy" attribute a definite priority over the attribute(s) of the single-proxy pruning algorithm. An alternative approach, that keeps the temporal locality nature of web access, is to allow the cache to prune a "Last-Copy" item in order to leave a high priority "Not-Last-Copy" item in the cache. A possible way to implement this idea is to determine a threshold on the merit scale of the "Not-Last-Copy" items list, above which items are not discarded from the "Not-Last-Copy" list but from the "Last-Copy" list. As an example, if a single-proxy pruning algorithm is LFU, the algorithm will prune an item from the "Not-Last-Copy" list only if this item was accessed in the last time more than

Δ seconds ago. If no more such items remain in the "Not-Last-Copy" list but more items have to be discarded from the cache, the algorithm starts discarding items from the "Last-Copy" list, provided of course that this list includes old items (that were accessed in the last time more than Δ seconds ago). However, this approach is not examined in this paper.

4   A Distributed Intra-cluster Scheme Based on the "Last-Copy" Concept

In this section we propose a distributed scheme for managing a cluster of caches while maintaining the "Last-Copy" attribute. This scheme guarantees with high probability the following properties:

1. Only one copy of each web item in the cluster is classified as "Last-Copy". An exception to this property might be if two requests for the same web item are received "almost simultaneously" by two different proxies when a "Last-Copy" for this particular web item does not exist in the cluster. In such a case, both proxies will mark their copy as "Last-Copy". However, as shown later one of these proxies will re-mark its copy as "Not-Last-Copy" after the copy is requested by a third proxy.
2. If one or more copies of a web item exist in the cache, a request for this item will be served by the local cluster rather than by the original server.

One may consider two models for the management of a cluster with the "Last-Copy" attribute:

1. A centralized model, where all the cluster participants are controlled by a centralized node that guarantees properties 1-2 above.
2. A distributed model, where no centralized controller exists. In such a case the cluster proxies need to run some intra-cluster protocol in order to maintain properties 1-2 above.

Although the implementation of a centralized model is simpler, we concentrate in this paper on the distributed approach. This is because the considered loosely-coupled system with a Layer-4 director does not scale well with a centralized controller that may easily become a bottleneck of the cluster. The work such a centralized controller has to perform, is much heavier than the "standard" work performed by a "regular" Layer-4 director.

Recall that in the considered system the proxy servers are connected to each other through a high-speed LAN, like 1Gb/s switched-Ethernet. Therefore IP multicast can be efficiently implemented within the cluster. The proposed scheme is based on the ICP (Internet Caching Protocol) as defined in [14]. ICP is a lightweight protocol, used for inter-proxy communication. The proxies exchange ICP query and reply messages in order to select the most appropriate neighboring proxy from which a requested object can be retrieved. A cache proxy sends an ICP query message, using either unicast or multicast, over UDP to its neighbors, and gets back from each neighbor a HIT or a MISS reply. The proxy analyzes the received reply messages and determines how to proceed accordingly. Related issues, like when to send an ICP query, to which cache proxy should an ICP query be sent, and how to process the received replies, are out of the scope of ICP.

In the proposed scheme, when a proxy receives a request for a web item that is not found in the local cache, it multicasts an ICP query message to the group of proxies in the cluster. One important advantage of our scheme over schemes that use ICP with multicast [12, 13] is that even if the searched item is found by many proxies, only one reply is sent back to the searching proxy. This reply is sent by the holder of the copy marked as "Last-Copy". Note that by minimizing the number of responses the searching proxy receives, we reduce the processing power this proxy needs to spend for opening and discarding irrelevant responses

Suppose that a proxy server p_i receives a user request for a web item w. The following cases are possible:

1. w exists in the local cache and it is valid.
2. w exists in the local cache but it is not valid.
3. w does not exist in the local cache.

Note: The proposed scheme is orthogonal to the validation model employed between the cluster proxies and the original web servers.
The response of p_i to the user request is as follows:

1. If w exists in the local cache and it is valid, it is sent to the requesting user.
2. If w exists in the local cache but it is not valid then:

   (a) If $p_i(w)$, namely the copy of w at p_i, is marked as "Last-Copy", then p_i checks the validity of $p_i(w)$ against the original server. If $p_i(w)$ is not fresh, p_i gets a fresh copy of w from the original server and sends it to the requesting user. The status of the fresh copy remains "Last-Copy".
   (b) If $p_i(w)$ is not marked as "Last-Copy", then w is requested from the local proxy that maintains the "Last-Copy" of w. Let this proxy be p_j. Proxy p_i discovers p_j by multicasting a special ICP query message called "Last-Copy-Search". The protocol then continues as follows:
       i. If $p_j(w)$ is valid, then p_j sends w to p_i, and p_i sends it to the user.
       ii. If $p_j(w)$ is not valid, then p_j informs p_i by unicast that the "Last-Copy" is not valid. In addition p_j marks its copy as "not-Last-Copy". Proxy p_i requests a fresh copy to be cached from the original server, marks it as "Last-Copy", and sends it to the user.
       iii. If no reply from a "Last-Copy" owner of w is received within a time-out period, p_i requests a fresh copy to be cached from the original server, marks this copy as "Last-Copy", and sends it to the requesting user.

3. If w does not exist in the local cache, p_i sends a multicast "Last-Copy-Search" and continues as in step 2(b) above.

According to the scheme described so far, the proxy that marks its copy of some web item as "Last-Copy", is the proxy that has received this item from the origin server rather than from another proxy. Nevertheless, due to a premature timeout or to ICP message loss, it is possible that two or more copies of the same item are marked as "Last-Copy" at the same time. Suppose, for example, that a proxy p_i searches for an item w whose "Last-Copy" version exists at proxy p_j. If p_j does not respond before p_i times out, or if the response message of p_j is lost, p_i will bring a copy of w from the original server and mark it as "Last-Copy".

However, the existence of two or more copies marked as "Last-Copy" does not affect the correctness of the whole scheme. Moreover, when a "Last-Copy-Search" message for this copy is multicast in the next time, this problem is reduced in the following way. Let the searching proxy be p_i, the responding proxies be p_j and p_k, and the web item be w. Let the IP addresses of these proxies be $Addr(p_i)$, $Addr(p_j)$ and $Addr(p_k)$ respectively. Note that p_i knows $Addr(p_j)$ and $Addr(p_k)$. If $Addr(p_j) < Addr(p_k)$, then p_i requests w from p_j and vice versa. The usage merit (LFU, LRU etc.) of w in p_k will be decreased, and it will eventually be pruned.

5   Simulation Results

In order to evaluate the performance improvement of the proposed scheme, we modeled a distributed proxy cluster managed by a Layer-4 director. The director uses a simple round-robin scheduling algorithm in order to assign incoming requests to the cluster proxies. We selected the two most popular single-proxy pruning algorithms, LRU and LFU, and evaluated the contribution of the "Last-Copy" scheme by comparing their performance, in terms of Object-Hit-Ratio and Byte-Hit-Ratio, with and without the "Last-Copy" scheme. When the "Last-Copy" scheme is not implemented, a proxy returns the requested web item if it exists in its local cache, and sends a query message to the cluster participants if the item does not exist. If no reply from one of the other proxies is received within a time-out period, the proxy requests a fresh copy from the original server, and returns it to the requesting user. If one or more replies are received, the proxy requests the web item from the first proxy to respond and forwards it to the requesting user. When the "Last-Copy" scheme is implemented, the proxy algorithm is as described in Section 4. In both cases a request received by a proxy is a "hit" if the requested object exists *somewhere* in the cluster. It is a "miss" if it does not exist anywhere in the cluster.

We used a ClarkNet-HTTP trace in order to test the behavior of the cluster. ClarkNet is a full Internet access provider for the Metro Baltimore-Washington DC area. The trace contains one-week worth of all HTTP requests[1]. Each request is associated with a timestamp taken from a clock with a granularity of 1 second. We created from this trace an input file containing only the successful

---

[1] The logs were collected by Stephen Balbach of ClarkNet, and contributed by Martin Arlitt and Carey Williamson of the University of Saskatchewan.

transactions, namely transactions to which an HTTP "200 OK" response is received. According to [2], 88.8% of the transactions (about 1,400,000) were found to be successful. Figure 1 depicts the size distribution of the items in the trace. It is evident that most of the items are relatively small (less than 10KB). In order to validate our simulation model, we started with a 1-proxy cluster that uses LRU pruning without the "Last-Copy" scheme, and found the results to be very similar to those reported in [5] for NLANR trace driven simulation.

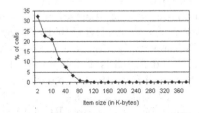

Fig. 1. Calls distribution according to item size

Figure 2 depicts the results achieved for a cluster of 15 proxies when LFU is used as the single-proxy pruning algorithm. In Fig. 2(a) we show the Object-Hit-Ratio, whereas in Fig. 2(b) we show the Byte-Hit-Ratio. The X-axis indicates the relative cache size of the *whole cluster*, namely the ratio between the cluster capacity and the total size of all the requested items. We see that the "Last-Copy" scheme increases the Object-Hit-Ratio and the Byte-Hit-Ratio substantially. It is interesting to note that the relative contribution of the scheme is only slightly affected by the cache size. We see an increase of 45% (from 55% to 80%) in the Object-Hit-Ratio when the relative cache size is 7%, and an increase of 40% when the relative cache size of the whole cluster is 20%.

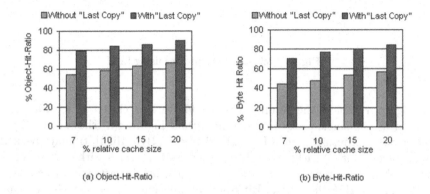

Fig. 2. A cluster of 15 proxies using LFU

Figure 3 depicts the results achieved for a cluster of 15 proxies when LRU is used as the single-proxy pruning algorithm. Generally the Byte-Hit-Ratio increases when large items are requested very often. Since in the considered trace, that reflects the situation in the Internet today, most of the popular items are relatively small (Fig. 1), the Byte-Hit-Ratio achieved for both LFU and LRU (Fig. 2(b) and Fig. 3(b) respectively) is smaller than the Object-Hit-Ratio (Fig. 2(a) and Fig. 3(a)). However, since the "Last-Copy" scheme increases the availability of less popular items, namely larger items in our trace, it is evident that for both LFU and LRU the contribution of "Last-Copy" to the Byte-Hit-Ratio is higher than to the Object-Hit-Ratio. For instance when the cache relative size is 7% the contribution of the "Last-Copy" scheme with LFU is 45% to the Object-Hit-Ratio and 55% to the Byte-Hit-Ratio. Similarly, for LRU the numbers are 14% and 20% respectively. However, the most important conclusion one may derive from Fig. 2 and Fig. 3 is as follows. Without the "Last-Copy" scheme, the Object-Hit-Ratio and Byte-Hit-Ratio performance of LRU are much better than those of LFU. With the "Last-Copy" scheme the differences almost disappear: LFU performs slightly better than LRU with regard to the Object-Hit-Ratio and the Byte-Hit-Ratio. This implies that the influence of LFU as a part of sophisticated web caching pruning algorithms, such as the Hybrid Algorithm mentioned in [17] and the LRV algorithm mentioned in [6], can achieve better Object-Hit-Ratio and the Byte-Hit-Ratio results in a cluster of proxies, when the "Last-Copy" scheme is employed.

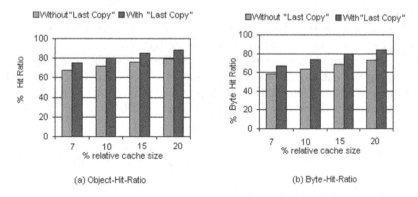

(a) Object-Hit-Ratio   (b) Byte-Hit-Ratio

Fig. 3. A cluster of 15 proxies using LRU

Figure 4 depicts the results achieved for various cluster sizes when LFU is used as the single-proxy pruning algorithm. Figure 4(a) shows the Object-Hit-Ratio improvement while Fig. 4(b) shows the Byte-Hit-Ratio improvement. As expected, for all cache sizes the improvement increases with the size of the cluster, because less non "Last-Copy" items are kept. For instance with a relative cache size of 15% the Object-Hit-Ratio improvement when the cluster has 5 proxies is 20%, while with 40 proxies it is 55%. Similarly the Byte-Hit-Ratio increases by 28% and 80% respectively. It is also evident that when the cluster is "relatively large", namely contains more than 10 proxies, the contribution of the "Last-Copy" scheme decreases as the cache size increases.

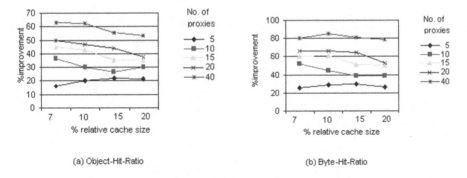

(a) Object-Hit-Ratio   (b) Byte-Hit-Ratio

Fig. 4. LFU improvement as a function of relative cache size and number of proxies

Figure 5 depicts the Object-Hit-Ratio and the Byte-Hit-Ratio as a function of the number of proxies in the cluster, when the relative cache size is 20% and LFU is used as the single-proxy pruning algorithm. The most interesting conclusion from these graphs is that without the "Last-Copy" scheme the Hit-Ratio decreases when the size of the cluster increases, whereas with the "Last-Copy" scheme the Hit-Ratio remains constant.

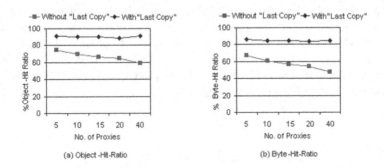

(a) Object -Hit-Ratio   (b) Byte -Hit-Ratio

Fig. 5. The Results for LFU when the relative cache size of the whole cluster is 20%

The same conclusion can be derived for LRU (Fig. 6). The explanation to this is as follows. Since we consider a "miss" each request that does not exist in the whole cluster, when the number of proxies increases and the "Last-Copy" scheme is not employed, there is a high probability for a popular item to be stored at multiple caches. This decreases, of course, the total number of items that can be cached in the whole cluster, and therefore reduces the Hit-Ratio. When the "Last-Copy" scheme is implemented with LFU or LRU, a lower pruning probability is assigned to items that do not exist in other caches of the cluster. Therefore the number of items stored in the whole cluster is not affected by the number of proxies in the cluster and the Hit-Ratio remains constant.

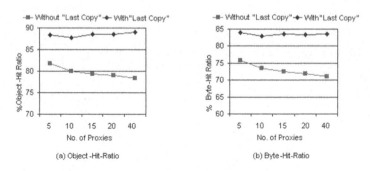

(a) Object-Hit-Ratio   (b) Byte-Hit-Ratio

Fig. 6. The Results for LRU when the relative cache size of the whole cluster is 20%

In the discussion so far we considered a "hit" each request for an object/byte that exists in any proxy of the cluster. However, in order to investigate a possible drawback of the proposed "Last-Copy" scheme, we need to distinguish between a "Local-Hit-Ratio" and a "Cluster-Hit-Ratio". The former term applies to the case where the request received by a proxy can be fulfilled using the proxy local cache only, whereas the latter term applies to the case where the request can be fulfilled using all the cluster caches. Each request that is a "Cluster-Hit" but not a "Local-Hit" requires the transfer of the object between two cluster proxies, and therefore increases the load on the cluster. Figure 7 depicts the "Local-Object-Hit-Ratio" of LRU and LFU for a cluster of 15 proxies. As expected, the "Local-Hit-Ratio" is lower when the "Last-Copy" scheme is employed. This is because the simulated scheme assigns a lower priority to "non-Last-Copy" items even if their LRU/LFU priority is high.

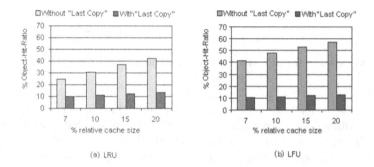

(a) LRU   (b) LFU

Fig. 7. "Local-Hit-Ratio" of a 15-proxy cluster with LRU and LFU

However, despite of the lower "Local-Hit-Ratio" of the "Last-Copy" scheme, the "Last-Copy"-based approach does not necessarily increase the processing load of the cluster proxies. Recall that when the "Last-Copy" approach is implemented, each ICP search message is responded by at most one proxy - the one that holds the "Last-Copy" version of the requested item. In contrast, when the "Last-Copy" approach is not implemented, each search message is responded by all the proxies of the cluster [12]. Hence, much more control messages have to be sent and processed. Figure 8 shows the number of control (ICP search and reply) messages sent with and without the "Last-Copy" scheme for a cluster of 15 proxies when LRU and LFU are used. Since according to Fig. 7 the "Local-Hit-Ratio" of both LFU and LRU while using the "Last-Copy" scheme is about 12%, the number of internal communication messages is also almost fixed: about 1.6 requests per message. When our scheme is not implemented, the "Local-Hit-Ratio" increases, the number of query control messages decreases, but since each proxy responds to every request message, the number of internal control messages per request increases substantially to 9-11 for LRU and 6.5-9 for LFU.

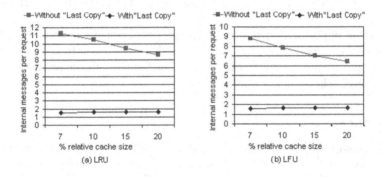

Fig. 8. Number of internal messages per request

6   Conclusions

In this paper we proposed an efficient scheme for increasing the cache hit-ratio in a loosely-coupled cluster, where proxies share cacheable content using some inter-cache communication protocol. The main part of the proposed scheme is an algorithm that increases the performance of any cache pruning algorithm when the latter is implemented in a cluster of cache servers. The solution presented in this paper is based on the concept of "Last-Copy" policy. According to this policy, exactly one copy of each web item in the cluster is marked as "Last-Copy". Items marked as "Last-Copy" are assigned a much lower pruning priority than "non-Last-Copy" items. This concept increases the number of web items available at the cluster, and therefore increases the cache hit ratios. The paper proposed an ICP-based cache sharing scheme that ensures that indeed only one copy of each item is marked as "Last-Copy".

In order to evaluate the performance improvement of the proposed scheme, we modeled a distributed proxy cluster managed by a Layer-4 director. The director employs a simple round-robin scheduling algorithm in order to assign incoming requests to the cluster proxies. We evaluated the contribution of the "Last-Copy" scheme by investigating the performance of the two most popular single-proxy pruning algorithms: LFU and LRU, with and without the "Last-Copy" scheme.

The simulations show that the "Last-Copy" scheme increases the Object-Hit-Ratio and the Byte-Hit-Ratio substantially. The relative contribution of the scheme is only slightly affected by the cache size. Since the scheme increases the availability of less popular items, namely larger items in our trace, it is evident that for both LFU and LRU the contribution of "Last-Copy" to the Byte-Hit-Ratio is higher than its contribution to the Object-Hit-Ratio. Another interesting result is that without the "Last-Copy" scheme, the Object-Hit-Ratio and Byte-Hit-Ratio of LRU are much better than for LFU. However with the "Last-Copy" scheme the differences disappear and both algorithms perform very similarly.

Another important advantage of the proposed scheme is that it decreases the internal communication substantially. This is because only the owner of the "Last-Copy" item responds to each request, rather than all the proxies when the "Last-Copy" scheme is not implemented.

References

1. M. Abrams, C. R. Standridge, G. Abdulla, S. Williams, and E. A. Fox. Caching proxies: Limitations and Potentials. In *1995 World Wide Web Conference*, 1995.
2. M. Arlitt and C. Williamson. Web server workload characterization: The search for invariants. In *ACM SIGMETRICS*, Philadelphia, PA, USA, Apr. 1996.
3. I. Cooper, I. Melve, and G. Tomlinson. Internet Web Replication and Caching Taxonomy. RFC-3040, Jan. 2001.
4. L. Fan, P. Cao, , and J. A. A. Z. Broder. Summary Cache: A Scalable Wide-Area Web Cache Sharing Protocol. In *ACM SIGCOMM*, Vancouver, Canada, 1998.
5. K. Psounis and B. Prabhakar. A Randomized Web-Cache Replacement Scheme. In *IEEE Infocom 2001 Conference*, Apr. 2001.
6. L. Rizzo and L. Vicisano. Replacement Policies for a proxy cache. Technical Report RN/98/13, University College London, Department of Computer Science, Feb. 1998.
7. Rousskov and D. Wessels. Cache Digests. In *3rd International WWW Caching Workshop*, June 1998.
8. Y. Shi, E. Watson, and Y. Chen. Model-Driven simulation of world wide web cache policies. In *Winter Simulation Conference*, Dec. 1997.
9. P. Srisuresh and D. Gan. Load Sharing using IP Network Address Translation (LSNAT). RFC-2391, Aug. 1998.
10. R. Tewari, H. M. Vin, Asit, and D. Sitaramy. Resource-based Caching for Web Servers. In *SPIE/ACM Conference on Multimedia Computing and Networking*, Jan. 1998.

11. P. Vixie and D. Wessels. Hyper Text Caching Protocol (HTCP/0.0). RFC-2756, Jan. 2000.
12. D. Wessels and K. Claffy. Application of the Internet Cache Protocol (ICP). RFC-2187, Sept. 1997.
13. D. Wessels and K. Claffy. ICP and the Squid Web Cache, Aug. 1997.
14. D. Wessels and K. Claffy. Internet Cache Protocol (ICP). RFC-2186, Sept. 1997.
15. S. Williams, M. Abrams, C. Standridge, G. Abdulla, and E. Fox. Removal Policies in Network Caches for World-Wide Web Documents. In *SIGCOMM*, 1996.
16. S. Williams, M. Abrams, C. Standridge, G. Abdulla, and E. Fox. Errata for Removal Policies in Network Caches for World-Wide Web Documents, Feb. 1997.
17. R. P. Wooster and M. Abrams. Proxy Caching that estimates page load delays. In *6th International World-Wide Web Conference*, Santa Clara, California, USA, 1997.

Modeling Short-Lived TCP Connections
with Open Multiclass Queuing Networks*

M. Garetto, R. Lo Cigno, M. Meo, E. Alessio, and M. Ajmone Marsan

Dipartimento di Elettronica – Politecnico di Torino,
Corso Duca degli Abruzzi, 24 – 10129 Torino, Italy,
{garetto,locigno,michela,ajmone}@polito.it

Abstract. In this paper we develop an open multiclass queuing network model to describe the behavior of short-lived TCP connections sharing a common IP network for the transfer of TCP segments. The queuing network model is paired with a simple model of the IP network, and the two models are solved through an iterative procedure. The combined model needs as inputs only the primitive network parameters, and produces estimates of the packet loss probability, the round trip time, the TCP connection throughput, and of the average TCP connection completion time (that is, of the average time necessary to transfer a file with given size over a TCP connection). The model presentation is centered on TCP-Tahoe, but the model of TCP-Reno is also available and results are presented. The analytical performance predictions are validated against detailed simulation experiments in a realistic networking scenario, proving that the proposed modeling approach is accurate.

1 Introduction

Modeling the TCP behavior is receiving great attention both from the academic and the industrial community. The reason for such enormous interest is essentially one: The behavior of TCP drives the performance of the Internet. Over 90% of the Internet connections use TCP, and over 95% of the bytes that travel over the Internet use TCP; hence, the availability of an accurate model of TCP is a necessity for the design and planning of Internet segments and corporate Intranets.

Models appeared so far in the literature can be grouped in two classes:

1. Models that assume that the RTT and the loss characteristics of the IP network are known, and try to derive from them the throughput (and the delay) of TCP connections; to this class belong works as [1,2,3];
2. Models that assume that only the primitive network parameters (topology, number of users, data rates, propagation delays, buffer sizes, etc.) are known, and try to derive from them the throughput and the delay of TCP connections, as well as the RTT and the loss characteristics of the IP network; to this second class belong the works in references [4,5,6,7,8] and many others.

---

* This work was supported by the Italian Ministry for University and Scientific Research through the PLANET-IP Project.

G. Carle and M. Zitterbart (Eds.): PfHSN 2002, LNCS 2334, pp. 100–116, 2002.

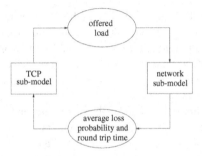

Fig. 1. High-level description of the model solution

Often, models in the second class incorporate a model similar to those of the first class, together with a simplified model of the network that carries the TCP segments and the two models are jointly solved through a fixed point iterative procedure. This situation is illustrated by Fig. 1, where the 'TCP sub-model,' describes the behavior of TCP connections, and the 'network sub-model,' focuses on the underlying IP network. The TCP sub-model computes the offered load to the system, given the average RTT and loss probability. The network sub-model estimates the RTT and the loss probability, given the offered load to the system.

With reference to our previous work, the contribution of this paper lies in the detailed presentation of the Open Multiclass Queueing Network method applied to TCP. The work presented in [7] assumed greedy connections using a closed queueing model. The work presented in [9] was concerned with a performance comparison between TCP-Tahoe and TCP-NewReno; results were obtained with OMQN models; however, the models themselves were not discussed. In [10], finally, we presented an overview of the modeling technique presented here, but a detailed application of the OMQN technique to a specific protocol, describing service times derivation and queues transition probabilities was not discussed in detail.

In this paper we use open multiclass queuing networks to develop a model that describes the behavior of an arbitrarily large number of short-lived TCP connections sharing a common infrastructure for the transfer of TCP segments. The model needs as inputs only primitive network parameters, and produces estimates of the packet loss probability, the RTT, the TCP connection throughput, and of the average TCP connection completion time. The TCP model presented here allows an unprecedented modeling accuracy for interacting short-lived TCP connections, yet maintaining a computationally simple solution. The level of insight in TCP dynamics offered by the model is also very high.

We use simulation experiments to validate our modeling approach, showing that the performance estimates that our models can generate are extremely accurate.

2   Queuing Network Models of TCP and Customer Classes

The TCP model in this paper is based on a description of TCP with an open multi-class queuing network (OMQN), in which all queues are of type $M/G/\infty$. Each queue represents a state of the TCP protocol, and each customer represents an active TCP connection. Classes are used to identify TCP connections according to the number of remaining segment transmissions before completion. Connections are opened when new customers enter the OMQN, then are used to transfer a file composed of N_P packets, and finally, after the file has been successfully delivered, the connection is closed as the customer leaves the OMQN.

Since all TCP connections are opened in the same state, only the queue corresponding to such state receives external arrivals. Let this queue be q^*.

Let $\lambda_{q,c}$ and $\lambda_{q,c}^e$ respectively denote the total and the external arrival rates at queue q in class c, and let $N_P^{\max}$ denote the maximum allowed file size in packets.

The external arrival rate of the OMQN is defined by the vector $[\lambda_{q^*,c}^e] = \lambda^e[\gamma_c]$ $1 \le c \le N_P^{\max}$, where λ^e is a traffic scaling factor, and $[\gamma_c]$ is a probability distribution vector describing the TCP connection file size.

Every time a new TCP segment is successfully transmitted, the class c of the customer is decreased by one; when the last packet is successfully transmitted (and the corresponding ACK is received), the customer leaves the OMQN, and the TCP connection closes.

The service times of the $M/G/\infty$ queues are independent from customer classes. All queues have an infinite number of servers, since there are no limitations to the number of TCP connections in any given state within the system.

From the specification of external arrival rates and service times, it is possible to derive the distribution of customers in the different queues and classes. From it we can derive the average number of customers of each class in every queue, $E[N_{q,c}]$, and the average TCP connection completion time.

The solution of the OMQN model is obtained solving the system of flow balance equations:

$$\lambda_{q,c} = \lambda_{q,c}^e + \sum_{i \in S} \sum_{j=c}^{N_P^{\max}} \lambda_{i,j} P(i,j;q,c) \qquad \forall(q,c) \qquad (1)$$

where S is the set of queues in the OMQN, and $P(i,j;q,c)$ is the transition probability from queue i in class j to queue q in class c. Note that $\lambda_{q,c}^e = 0$ for any $q \ne q^*$.

The complete model solution is obtained by iterating the solutions of the network sub-model and the TCP sub-model (as shown in Fig. 1) with a fixed point algorithm (FPA), until convergence is reached, according to a specified threshold. All the performance figures of interest depend on the average distribution of customers in the queues, which is obtained by solving the flow balance problem defined by equation (1). The complexity of this step is $O([M_q \cdot N_P^{\max}]^3)$. Moreover, exploiting the triangular structure of the system of linear equations, deriving from the fact that customers can only decrease their class (or keep it constant) on transitions, the complexity reduces to $O([M_q \cdot N_P^{\max}]^2)$. Since M_q is of the order of few hundreds, the CPU time required for each step of the fixed point algorithm is extremely small.

The number of iterations before convergence depends on the accuracy required of the FPA, but a relative accuracy of 10^{-6} is generally reached in a few tens of iterations.

3   The Model of TCP-Tahoe

Let W be the maximum TCP window size expressed in segments and C the maximum number of retransmissions before TCP closes the connection. Fig. 2 reports the OMQN model of short-lived TCP-Tahoe connections, in the case $W = 10$, which is the smallest maximum window size that allows a complete description of all the states of the proto-col. New TCP connections open in the state described by queue FE_1 (thus, $q^* = FE_1$). The shaded queues describe the steady-state behavior of greedy TCP connections, while the other queues describe the transient behavior of the protocol. Non shaded queues are needed to describe the first slow start phase, which is different from the others, since *ssthresh*, the threshold separating the slow start phase from the congestion avoidance phase, is not set, and the congestion window can grow exponentially to W. When trans-ferring short files, the initial transient has a major impact on performance. Customers leave the system whenever they reach class 0, regardless from the queue they are in; these transitions are not shown in Fig. 2 to avoid cluttering the graph.

The queues in Fig. 2 are arranged in a matrix pattern: all queues in the same row correspond to similar protocol states, and all queues in the same column correspond to equal window size. It is clear that the OMQN in Fig. 2 is highly structured, since, in spite of the large number of queues present in the model, all queues can be grouped in only 13 classes, described in some detail below, assuming that the reader is familiar with TCP-Tahoe (see [12] and [13]). Table 1 reports the service rates of the queues.

FE_i $(1 \leq i \leq W)$ **and** E_i $(1 \leq i \leq W/2)$ model the exponential window growth during slow start; the index i indicates the transmission window size. Queues FE_i are
  visited only during the first slow start phase after the connection is opened, while
  queues E_i model all other slow start phases; during the first slow start phase the
  growth is limited only by losses, since *ssthresh* is not yet assigned. As can be seen
  from Table 1, the average service time at these queues depends on the round-trip
  time through a factor σ which is an apportioning coefficient that takes into account
  the fact that during slow start the TCP transmission window grows geometrically
  with base 2 (see [7] for details); we use $\sigma = 2/3$.
ET_i $(1 \leq i \leq W)$ model the TCP transmitter state after a loss occurred during slow
  start: the transmission window has not yet been reduced, but the transmitter is
  blocked, because its window is full; the combination of window size and loss pattern
  forbids a fast retransmit (i.e., less than 3 duplicate ACKs are received). The TCP
  source is waiting for a timeout to expire. Queues ET_i $(W/2 \leq i \leq W)$ can be
  reached only from queues FE_i, i.e., during the first slow start. The service time is a
  function of the term T_0 which in the model is approximated with $\max(3\,\mathrm{tic}, 4\,\overline{\mathrm{RTT}})$,
  since the estimation of RTT variance is not available.
FF_i $(4 \leq i \leq W)$ **and** EF_i $(4 \leq i \leq W/2)$ model a situation similar to that of queues
  ET_i, where, instead of waiting for the timeout expiration, the TCP source is waiting
  for the duplicated ACKs that trigger the fast retransmit. Different queues have been
  used to distinguish the first loss event during a connection, i.e., fast retransmit events
  triggered from queues FE_i, from all others cases.
L_i $(2 \leq i \leq W)$ model the linear growth during congestion avoidance (notice that
  queue L_1 does not exist).

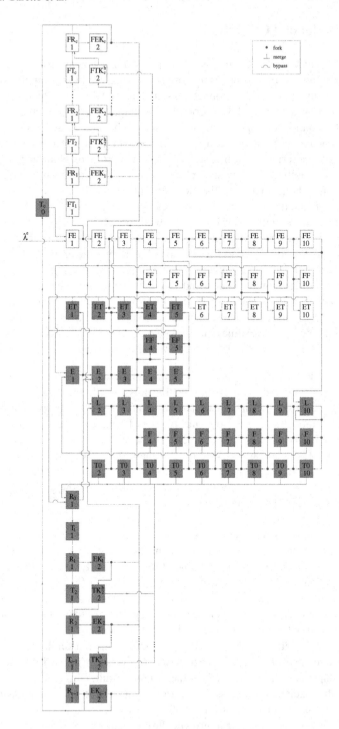

Fig. 2. The OMQN model of TCP-Tahoe

Table 1. Queues service times

Queue	Service time
E_1, FE_1	$\overline{\text{RTT}}$
E_2, FE_2	$\overline{\text{RTT}}$
$E_i, FE_i \quad i = 2^n \quad n \geq 2,$	$\sigma\,\overline{\text{RTT}}$
$E_i, FE_i \quad 2^{n-1} + 1 \leq i \leq 2^n - 1$	$\frac{(1-\sigma)\overline{\text{RTT}}}{2^{n-1}-1}$
$ET_i \quad 1 \leq i \leq 3$	$T_0 - \overline{\text{RTT}}$ [1]
$ET_i \quad i \geq 4$	T_0 [1]
EF_i	T_0 [1]
FF_i, F_i	$\frac{\overline{\text{RTT}}}{2} + \frac{4\overline{\text{RTT}}}{i}$
L_i	$\overline{\text{RTT}}$
$T0_2$	$T_0 - \overline{\text{RTT}}$ [1]
$T0_i \quad 3 \leq i \leq W$	$T_0 - \overline{\text{RTT}}/2$ [1]
$T_i, FT_i \quad 1 \leq i \leq 6$	$2^i\,T_0$ [1]
$T_i, FT_i \quad 7 \leq i \leq C - 1$	$64\,T_0$ [1]
R_i, FR_i, EK_i, FEK_i	$\overline{\text{RTT}}$
TK_i^1, FTK_i^1	$T_i - \overline{\text{RTT}}$ [1]
TK_i^2, FTK_i^2	$T_0 - \overline{\text{RTT}}$ [1]
T_C	$180\,\text{s}$

[1] Remember that T_0 and T_i assume different values
whether the first packet of a connection is lost or otherwise

$F_i\,(4 \leq i \leq W)$ model losses that trigger a fast retransmit during congestion avoidance.

$T0_i\,(2 \leq i \leq W)$ model the detection of losses by timeout during congestion avoidance.

$FT_i\,(1 \leq i \leq C)$ and $T_j\,(1 \leq j \leq C - 1)$ model the time lapse before the expiration of the i-th or $(j + 1)$–th timeout for the same segment, i.e., model the backoff timeouts. Queues FT_i model the backoff procedure in case the first segment is lost, when the RTT estimate is not yet available to the protocol and T_0 is set to a default value, typically 12 tics. This event, that may seem highly unlikely, has indeed a deep impact on the TCP connection duration.

Queue T_C models TCP connection that were closed for an excessive number of timeouts. Closed connections should leave the system; however, the model solution we use is correct only if all customers entering the OMQN in any class, leave in class 0 (a 'work conservation' principle), hence from queue T_C connections are supposed to re-open.

$FR_i\,(0 \leq i \leq C)$ and $R_j\,(0 \leq j \leq C - 1)$ model the retransmission of a packet when the timeout expires.

$FEK_i\,(0 \leq i \leq C)$ and $EK_j\,(1 \leq j \leq C - 1)$ model the first stage of the slow start phase (i.e., the transmission of the first 2 non-retransmitted packets) after a backoff timeout. During this phase the Karn algorithm ([14] and [12] Ch. 21.3) has a deep impact on the protocol performance under heavy load.

FTK_i^h $(0 \leq h \leq C)$ **and** TK_j^h $(1 \leq j \leq C - 1)$; **h=1,2**

model the wait for timeout expiration when losses occurred in queues FEK_i (EK_j); the superscript h in Fig. 2 discriminates two queues that are drawn together, but have different service times. The one with $h = 1$ is entered when the first packet of the pair transmitted in queues FEK_i (EK_j) is lost, the other when the lost packet is the second. The transition from FTK_i^1 (TK_i^1) is to queue FR_i (R_i), while the transition from FTK_i^2 (TK_i^2) is to queue FR_0 (R_0).

The average service time represents the time a typical connection spends in a given protocol state. The service times derive directly from the protocol properties and are mainly function of the average round-trip time $\overline{RTT}$. Service times are independent from the customer class, as can be seen in Table 1.

The transition probabilities between queues derive from the TCP dynamics and characteristics, and they are related mainly to the packet loss probability. It is well known (see for instance [2,7]), that TCP dynamics introduce correlations in packet losses, and these have a deep impact on performance and consequently on any modeling process. Indeed, with interacting TCP connections, there exist both short-term and long-term correlations. Short-term correlations are between packets within the same transmission window. On a temporal axis, this correlation extends for roughly a round trip time. Long-term correlations are much more difficult to identify, because they are generated by the interaction among different connections. Strictly speaking, an OMQN model cannot take into account long-term correlations, but some heuristics allows the approximation of its effect, by introducing the following quantities:

P_L – average packet loss ratio, $P_S = 1 - P_L$;

P_{L_f} – loss ratio for the *first* packet in a window; this is also the probability of loosing a burst of packets, $P_{S_f} = 1 - P_{L_f}$;

P_{L_a} – packet loss ratio *after* the first packet in a window is lost; this is also the packet loss ratio within a burst and takes into account the short term correlation, $P_{S_a} = 1 - P_{L_a}$;

$P_{L_{fc}}$ – loss ratio for the *first* packet in a window under long term *correlation*, $P_{S_{fc}} = 1 - P_{L_{fc}}$;

$P_T(i)$ – probability that the exponential growth threshold *ssthresh* has value i; $P_C(i)$ cumulative distribution of $P_T(i)$;

$Z_{i,j}^T - = \frac{P_T(i)}{1 - P_C(j)}$; $Z_{i,j}^C = \frac{1 - P_C(i)}{1 - P_C(j)}$.

The relationship between P_L, P_{L_f}, P_{L_a} and $P_{L_{fc}}$, as well as their derivation and rationale are discussed in Section 4.2.

Table 2, reports the transition probabilities that model the succesful trnsmission of a whole window. The probability $P(q_i, c; q_j, c - k)$ reported in the table is the tramsition from source queue q_i (first column) to destination queue q_j (second column), decreasing the customer class of k units (third column) under the conditions indicated in the last column; $\wedge$ is the logical AND operator. Deterministic transitions are not reported. These probabilities are rather straightforward to compute. The only modeling problem arises in transitions from exponential to linear growth, where the value of *ssthresh* is crucial. Unfortunately, the value of *ssthresh* does not depend on the current state, but on the state

Table 2. Transition probabilities modeling successful transmission events; $P(\cdot)=P(q_i,c;q_j,c-k)$

q_i	q_j	k	$P(\cdot)$	Condition
FE_1	FE_2	1	P_{S_f}	
FE_i	FE_{i+1}	2	$P^2_{S_f}$	$2 \le i \le W-1$
FE_W	L_W	2	$P^2_{S_f}$	
FEK_i	FE_3	2	$P^2_{S_f}$	
E_1	E_2	1	$P_{S_{fc}}$	
E_i	E_{i+1}	2	$P^2_{S_{fc}} Z^C_{i+1,i}$	$2 \le i < W/2$
	L_i	2	$P^2_{S_{fc}} Z^T_{i,i}$	
L_i	L_{i+1}	i+1	$P^{i+1}_{S_f}$	$2 \le i \le W-1$
L_W	L_W	W	$P^W_{S_f}$	
R_0	E_2	1	$P_{S_{fc}}$	
EK_i	L_2	2	$P^2_{S_{fc}} Z^T_{2,2}$	
	E_3	2	$P^2_{S_{fc}} Z^C_{3,2}$	

when the last packet was lost. Explicitly taking into account the value of *ssthresh*, leads to the explosion of the number of queues needed, that are approximately $W/2 \times M_q$. We resorted to a stochastic description of the distribution of *ssthresh*, that is taken into account by the terms $Z^T_{i,j}$ and $Z^C_{i,j}$ (see [7] for details).

Due to the lack of space, we do not report here all the transition probabilities of the model, that can be found in [15].

3.1 Packet Generation and Network Load

A TCP connection in a given queue q and class c generates a number of packets $\Pi_{q,c}$, which is the minimum between the number of packets allowed by the protocol state, Π_q, and the remaining packets to be transmitted during the TCP session, $\Pi_{q,c} = \min(\Pi_q, c)$. It follows that, from each queue q of the OMQN model, the load offered to the network by a TCP connection is

$$\Lambda_q = \sum_{c=1}^{N_P^{\max}} \lambda_{q,c} \Pi_{q,c} \tag{2}$$

where $\lambda_{q,c}$ is the arrival rate at queue q in class c computed from the OMQN. The total load offered to the network is then

$$\Lambda = \sum_{i \in S} \Lambda_q. \tag{3}$$

To solve (2) and (3) we need to define Π_q for each $q \in S$.

Some queues represent states of the protocol in which no packet is generated: queues T_i and FT_i, which model the backoff timeouts, and queue ET_1, which stands for the timeout expiration when the only packet in the window has been lost.

One packet per service is generated in queues R_i and FR_i, which model the retransmission of a packet when timeout expires, and in queues FE_1 and E_1, where the TCP window size equals 1.

The generation of two packets in queues FE_i and E_i, with $i > 1$, is due to the exponential window size increase in slow-start mode. Two packets per service are generated also in queues EK_i and FEK_i, standing for the transmission of the first 2 non-retransmitted packets after a backoff timeout. Similarly, two packets are generated in queues TK_i^3 and FTK_i^3.

More complex is the derivation of the number of packets that the protocol allows to generate for each service in queues EF_i, ET_i and FF_i, for $i \geq 3$, since it depends on which packet was lost in the previously visited queue. It results: $\Pi_{EF_i} = \Pi_{ET_i} = \Pi_{FF_i} = \Pi_i$

$$\Pi_i = \left[(i - 2) \frac{P_{ll}}{1 - P_{ss}^2} + (i - 1) \frac{P_{ss} P_{ll}}{1 - P_{ss}^2} \right] \tag{4}$$

where P_{ll} (P_{ss}) assumes the values P_{L_f} or $P_{L_{fc}}$ (P_{S_f} or $P_{S_{fc}}$), depending on the previously visited queue (see Section 4.2). Instead, the load offered by the connections in queues L_i is:

$$\Pi_{L_i} = \left[P_{L_f} i + (1 - P_{L_f})(i + 1) \right] \quad i \leq W - 1 \tag{5}$$

$$\Pi_{L_W} = W \tag{6}$$

The computation of the load offered to the underlying IP network by connection at individual queues of type F_i and $T0_i$ is cumbersome, since for each queue it depends on the loss pattern. On the contrary, if we consider the aggregate load collectively offered by the set of queues

$$\Phi_{q_{ll}} = \bigcup_i [F_i \cup T0_i] \tag{7}$$

then its computation is much easier

$$\Lambda_{\Phi_{q_{ll}}} = \sum_{i=2}^{W} \left[\sum_{j=1}^{i} j \, P_{S_f}^j P_{L_f} \right] \lambda_{L_i} \tag{8}$$

Finally, for queues TK_i^2, FTK_i^2 and ET_2, we can write:

$$\Pi_{TK_i^2} = \Pi_{FTK_i^2} = \Pi_{ET_2} = \frac{P_{ss} P_{ll}}{P_{ll} + P_{ss} P_{ll}} . \tag{9}$$

4   Modeling the Underlying IP Network

Since the modeling focus in this paper is on TCP, the network model is kept as simple as possible. Nevertheless, it must allow the correct estimation of the key parameters that drive TCP performance. We assume that in its overall performance the network has a "drop tail" behavior.

A single server queue is the simplest possible model for the underlying IP network. The network is modeled as an $M^{[D]}/M/1/B$ queue, where packets arrive in batches whose size varies between 1 and W with distribution $[D]$. The variable size batches model the burstiness of the TCP transmission within the $\overline{RTT}$. The Markovian arrival of the batches finds its reason mainly in the Poisson assumption for the TCP connections arrival process, as well as in the fairly large number of connections present in the model. The Markovian assumption about service times is more difficult to justify; indeed, the transmission time of fixed length packets is constant, but the influence on results of using an $M^{[D]}/D/1/B$ queue was observed not to be worth the increased complexity.

4.1 Evaluation of $\overline{RTT}$ and P_L

The average packet loss ratio P_L, and the average queueing time t_B, are obtained directly from the solution of the $M^{[D]}/M/1/B$ queue; $\overline{RTT}$ is then estimated as the sum of t_B, the average two-way propagation delay of connections, and the packet transmission time in routers. The key point of the modeling process is the determination of the batch size distribution $[D]$. We compute the batch sizes starting from the number of segments N_R sent by a TCP transmitter during $\overline{RTT}$. This is clearly a gross approximation and a pessimistic assumption, since TCP segments are not transmitted together, and they do not arrive together at the router buffers. To balance this pessimistic approximation, the batch size is reduced by a factor $\mu < 1$. Actually, since the TCP burstiness is much higher during slow start than during congestion avoidance, we use two different factors: μ^e during the exponential window growth, and μ^l during the linear window growth. Unfortunately, it was not possible to find a direct method for the computation of these two factors, but a simple heuristic optimization led to the choice $\mu^e = 2/3$, $\mu^l = 1/3$, that yields satisfactory results in every tested scenario.

The computation of N_R is straightforward, starting from the solution of the OMQN of Fig.2. For every queue whose service time is $\overline{RTT}$, $N_R = \Pi_q$. In the other cases, queues must be grouped, based on the protocol dynamics. For instance, during exponential growth the grouping is based on the window doubling.

Besides the computation of batch sizes, it is also necessary to evaluate the generation rates of batches. For reasons that will be clear in Section 4.2, we separate the generation of bursts during the first slow start phase and during congestion avoidance, the relative rate being $\lambda_{bf}(i)$, from the generation of bursts during steady-state slow start phases, whose rate is $\lambda_{bc}(i)$. $\lambda_{bf}(i)$ and $\lambda_{bc}(i)$ are different for each session length, since, specially for short sessions, the number of packets to be transferred influences the probability that a connection visits a given queue (e.g., a connection that must transfers 10 packets will never transmit a packet with window size larger than 6).

4.2 Evaluation of Short- and Long-Term Correlations

Describing the TCP model in Section 3, we introduced the loss probabilities P_{L_f}, P_{L_a} and $P_{L_{fc}}$, that take into account short and long term correlations in packet losses. The short term correlation due to loss detection latency is well known, and it is easily accounted for in our model, since it extends only over a single window.

Long term correlation is due to a completely different phenomenon, rooted in the elastic nature of TCP traffic, that extends congestion over time.

In order to approximate this phenomenon, we assume that long term correlations extend in the slow start phase that follows a loss event: During this phase the probability of experiencing another loss event is higher, and it is represented by $P_{L_{fc}}$. Newly opened connections and connections in congestion avoidance phases experience instead a lower probability (P_{L_f}) of incurring in a loss event.

Assume that the loss correlation is positive, and is represented by multiplicative coefficients: $K_s > 1$ for the short term and $K_l > 1$ for the long term. The following equations allow the computation of P_{L_f}, P_{L_a} and $P_{L_{fc}}$ starting from the average loss probability P_L

$$P_{L_a} = K_s P_{L_f} \; ; \; P_{L_{fc}} = K_l P_{L_f}$$
$$\Lambda P_L = \Lambda P_{L_a} + (1 - K_s) \sum_{i=1}^{W} \lambda_{bf}(i)[1 - (1 - P_{L_f})^i]$$
$$+ (1 - \frac{K_s}{K_l}) \sum_{i=1}^{W} \lambda_{bc}(i)[1 - (1 - P_{L_{fc}})^i].$$

(10)

Unfortunately, the coefficients K_s and K_l depend on the network congestion, and are far from easy to compute, hence we need some simplifying assumptions. First of all, let's assume for simplicity $P_{L_{fc}} = P_L$, i.e., the probability of losing a burst of packets during a steady-state slow start phase is equal to the average packet loss ratio. If short term correlations are stronger than long term ones, as it is reasonable, we also have $P_{L_f} < P_{L_{fc}} < P_{L_a}$, so that our assumption that $P_{L_{fc}} = P_L$ is also equivalent to assuming that the average loss ratio is equal to the median value of the approximating discrete distribution.

The works in [2] and [7], hints to the possibility that the loss probability within the window of the first lost packet is almost constant. The value that seems to best fit a general scenario is $P_{L_a} = 0.2$. Once P_{L_a} and $P_{L_{fc}}$ are determined, P_{L_f} is computed from Eqs. (10).

5 Computation of the Session Duration

In order to compute the average session duration, let's assume that all customers enter the OMQN with the same class N_P. The average time Θ spent in the OMQN is derived from Little's theorem $\Theta = \frac{N}{\lambda^e}$, where N is the total average number of customers in the OMQN, and λ^e is the external arrival rate of customers.

When the initial class of the arriving customers is not constant, Little's result computes the average over all connections, regardless of the initial class. Class N_P customers comprise connections entering the OMQN with class N_P as well as all those connections whose initial class is larger than N_P, but still have N_P packets to transmit. Yet, the traffic mix influences the overall network performance, hence we devised a solution based on a two-step solution of the OMQN model.

Step 1 – The model is solved with the chosen external arrival distribution $\lambda^e \; [\gamma_c]$. During this step, parameters such as $\overline{\text{RTT}}$, P_L, K_s and K_l are computed.

Step 2 – Using the parameters computed in step 1, the OMQN model is solved again for each external arrival class $c = 1, \ldots N_P^{\max}$. During this step the *ssthresh* distribution is computed again, since the one obtained in step 1 is not representative of each input class (e.g., a connection with 10 packets to transmit cannot have *ssthresh* = 20), but only of their average. For the same reason $\lambda_{bf}(c)$, $\lambda_{bc}(c)$, P_{L_f} and P_{L_a} are also recomputed. Finally, Little's result is used to compute each input class latency.

6 Validation and Results

In order to validate our analytical model of TCP, we compare the performance predictions obtained from the OMQN model solution against point estimates and 95% confidence intervals obtained from very detailed simulation experiments. The tool used for simulation experiments is *ns version 2* [11]; confidence intervals were obtained with the "batch means" technique, using 30 batches. Simulations last for 1,000–2,000 s when the bottleneck is a 45 Mbit/s link, and are four times longer when the bottleneck is a 10 Mbit/s link.

6.1 Network Topology and Traffic Load

We chose a networking environment which closely resembles the actual path followed by Internet connections from our University LAN to Web sites in Europe and the USA, where two clearly distinct traffic patterns can be identified.

The topology of the network we consider is shown in Fig. 3; at the far left we can see a set of terminals connected to the internal LAN of Politecnico di Torino. These terminals are the clients of the TCP connections we are interested in (white circles in the figure represent TCP clients; grey circles represent TCP servers). The distance of these clients from the Politecnico router is assumed to be uniformly distributed between 1 and 10 km. The LAN of Politecnico is connected to the Italian research network, named GARR-B/TEN-155, through a 10 Mb/s link whose length is roughly 50 km (this link will be called POLI-TO). Internally, the GARR-B/TEN-155 network comprises a number of routers and 155 Mb/s links. One of those connects the router in Torino

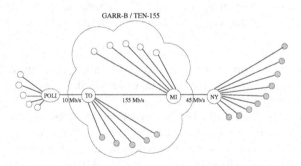

Fig. 3. Abstract view of the Internet from the Politecnico di Torino LAN; servers are shaded, clients are white

with the router in Milano; its length is set to 100 km. Through the GARR-B/TEN-155 network, clients at Politecnico can access a number of servers, whose distance from Politecnico di Torino is assumed to be uniformly distributed between 100 and 9,900 km. From Milano, a 45 Mb/s undersea channel whose length is about 5,000 km reaches New York, and connects GARR-B to North-American Internet backbones (this link will be called MI-NY). Many other clients use the router in Milano to reach servers in the US. The distance of those clients from Milano is assumed to be uniformly distributed between 200 and 2,800 km. The distance of servers in the US from the router in NY is assumed to be uniformly distributed between 200 and 3,800 km.

At present, the GARR-B/TEN-155 network is over-dimensioned, thus it is never congested. Both the POLI-TO and the MI-NY link, instead, are bottleneck channels that often incur heavy and persistent congestion. From the abstract network represented in Fig. 3 we carve out the following two different traffic patterns that give rise to the two scenarios we name the US BROW and the LOC ACC.

US Brow: This scenario becomes relevant when the traffic pattern makes the MI-NY link the bottleneck of the system. When this is the case, we have connections with lengths between 5,400 and 11,600 km that compete for the 45 Mbit/s link.

Loc Acc: This scenario corresponds to moments when the access link of our University is the bottleneck of the system. Connections compete for the 10 Mbit/s on the POLI-TO link, and their lengths are distributed between 151 and 9,960 km.

The packet size is 1,024 bytes; the maximum window size is 64 packets. We consider the cases of buffer sizes equal to either 128 or 64 packets, and TCP tic equal to 500 ms. The amount of data to be transferred by each connection (i.e., the file size) is expressed in number of segments: 50% of the connections are 10 segments long 40% are 20 and 10% are 100.

6.2   Numerical Results

Fig. 4 reports the average packet loss probability computed with the model, as well as point estimates and confidence intervals obtained via simulation in both the US BROW and LOC ACC scenarios. Results are plotted versus the normalized external load, which is the load the network would have if no packets were lost, so that no retransmissions are necessary. The upper curve refers to the case of 64 packet buffers, while the lower one refers to the case of 128 packet buffers. Markers correspond to simulations and report the confidence intervals.

As predicted by the model, the different conditions of the US BROW and LOC ACC scenarios have a minor impact on the average packet loss probability, that is predicted very accurately by the model over a large range of network loads.

Fig. 5 reports the average completion time for 10 segments files (left plot) and 100 segments files (right plot). The file size has a major impact on results (notice the different y-scales of the plots), as expected. The TCP performance is still dominated by the buffer size (that drives the loss ratio), while the scenario has a minor impact.

The presence of instability asymptotes around load 0.97 is evident in both figures. The vertical asymptotes correspond to points where the actual load of the network (considering retransmissions) approaches 1.

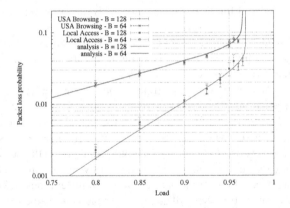

Fig. 4. Packet loss probability as a function of the external load

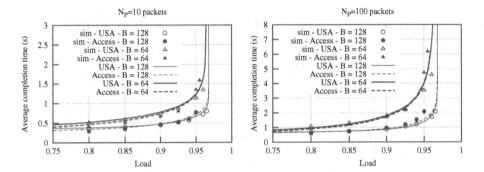

Fig. 5. Average session completion time as a function of the external load; 10 packets sessions in the left plot and 100 packets sessions in the right plot

6.3   Extension to Other TCP Versions

Up to this point we have only described or modeling approach in the case of TCP-Tahoe, but our method can be extended to other TCP versions as well. So far we have fully modeled TCP-NewReno, whose model is also briefly discussed in [8], though that paper focuses more on modeling a multibottlenek IP network.

The main difference between Tahoe and NewReno lies in the presence of the fast recovery procedure [13], which avoids the slow start phase if losses are detected via duplicated ACKs, entering congestion avoidance after halving the congestion window. The number of queues and queue types needed for the description of TCP-NewReno, is approximately the same as for TCP-Tahoe, though some queues change their meaning, for instance to model the fast recovery procedure where the actual transmission window keeps growing to allow the transmission of new packets while the lost ones are recovered.

Lack of space forbids a detailed description of this model, giving all the queues service times and transition probabilities. Instead, we present numerical results for the same cases we considered for TCP-Tahoe, again with validation results obtained with *ns-2*.

Fig. 6 reports the average packet loss probability in the same conditions as those of Fig. 4. The loss probability is similar to TCP-Tahoe, but the instability asymptotes are now closer to 1 (namely, a little beyond load 0.98).

For what concerns the session transfer time, we present the results in a different form, so as to give additional insight. Fig. 7 reports the results for buffer 128 only. the case with buffer 64 yields similar results, not reported for lack of space. The left plot refers to the US Brow scenario, while the right one refers to the Loc Acc scenario. Both plots report three curves, one for each type of connection mix in the network. For light and medium loads, the session latency is dominated by the transmission delay, since loss events are rare; hence longer sessions have longer transfer times and the increase is roughly linear. The interesting fact is that the asymptote is the same for all connection lengths, meaning that the protocol is reasonably fair towards connections of different lengths.

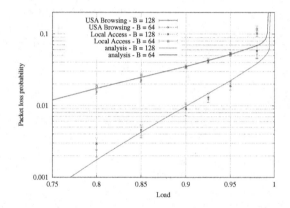

Fig. 6. Packet loss probability as a function of the external load for NewReno connections

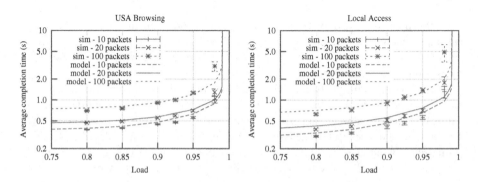

Fig. 7. Average session completion time as a function of the external load for the US Brow scenario (left plot) and for the Loc Acc scenario in the case of 128 packets buffer

The same is true also for TCP-Tahoe, but it is less evident from the presentation format of Fig. 5, which, on the other hand, better highlights the role of the buffer and, by contrast, the small role of the bottleneck capacity on the transfer delay.

7 Conclusions

In this paper we have developed an OMQN model to very accurately describe the behavior of short-lived TCP connections that exploit a common IP network for the transfer of finite-size files.

Starting from the primitive network parameters, the queuing network model is solved together with a simple model of the IP network, through an iterative procedure, and the average time for the completion of the file transfer is estimated, along with the TCP connection throughput and goodput.

The presented modeling approach was applied to TCP-Tahoe, and TCP-Reno (which is also shortly discussed in [8]), and it is currently being extended also to TCP-SACK.

The analytical performance predictions generated by the model for realistic networking scenarios have been validated against detailed simulation experiments, proving that the proposed modeling approach is extremely accurate and flexible, besides providing an excellent insight into TCP dynamics and behavior.

Results validating the model discussed in the paper were obtained for relatively slow networks (10–45 Mbit/s bottlenecks) due to the impossibility of obtaining reliable simulation results with higher speeds. It must be noted, however, that the model complexity is *independent* from the network speed. Hence OMQN models are a good means to ferecast the bahavior of protocols like TCP in very high speed netorks.

References

1. V. Paxson, "Empirically-Derived Analytic Models of Wide-Area TCP Connections," *IEEE/ACM Transactions on Networking*, 2(4):316–336, August 1994.
2. J. Padhye, V. Firoiu, D. Towsley, and J. Kurose, "Modeling TCP Throughput: A Simple Model and its Empirical Validation," *Proc. ACM SIGCOMM'98*, Sept., 1998.
3. N. Cardwell, S. Savage, T. Anderson, "Modeling TCP Latency," *Infocom 2000*, Tel Aviv, Israel, March 2000.
4. V. Mishra, W. B. Gong, D. Towsley, "Fluid-based Analysis of a Network of AQM Routers Supporting TCP Flows with and application to RED", *in Proc. SIGCOMM'2000*, Aug. 28–Sept. 1 2000, Stockholm, Sweden.
5. C. V. Hollot, V. Mishra, W. B. Gong, D. Towsley, "A Control Theoretic Analysis of RED," *in Proc. IEEE Infocom 2001*, Anchorage, Alaska, USA, April 22–26, 2001.
6. A. Kumar, "Comparative Performance Analysis of Versions of TCP in a Local Network with a Lossy Link," *IEEE/ACM Transactions on Networking*, 6(4):485–498, August 1998.
7. M. Garetto, R. Lo Cigno, M. Meo, M. Ajmone Marsan, "A Detailed and Accurate Closed Queueing Network Model of Many Interacting TCP Flows," *in Proc. IEEE Infocom 2001*, Anchorage, Alaska, USA, April 22–26, 2001.
8. M. Garetto, R. Lo Cigno, M. Meo, M. Ajmone Marsan, "Queuing Network Models for the Performance Analysis of Multibottleneck IP Networks Loaded by TCP Short Lived Connections," Politecnico di Torino Tech. Rep. DE/RLC/2001-5, Politecnico di Torino, June 2001. Available at http://www.tlc-networks.polito.it/locigno/papers/de-rlc-01-5.ps

9. E. Alessio, M. Garetto, R. Lo Cigno, M. Meo, M. Ajmone Marsan, "Analytical Estimation of the Completion Time of Mixed NewReno and Tahoe TCP Traffic over Single and Multiple Bottleneck Networks," *Proc. IEEE Globecom 2001*, San Antonio, Tx, USA, Nov. 25–29, 2001.

10. M. Garetto, R. Lo Cigno, M. Meo, M. Ajmone Marsan, "On the Use of Queueing Network Models to Predict the Performance of TCP Connections," *Proc. 2001 Tyrrhenian International Workshop on Digital Communications*, Taormina (CT), Italy, Sept. 17–20, 2001.

11. ns-2, network simulator (ver.2). LBL, http://www-mash.cs.berkeley.edu/ns.

12. W. R. Stevens. *TCP/IP Illustrated, vol. 1*. Addison Wesley, Reading, MA, USA, 1994.

13. W. Stevens, "TCP Slow Start, Congestion Avoidance, Fast Retransmit, and Fast Recovery Algorithms," RFC 2001, IETF, Jan. 1997

14. P. Karn, C. Partridge, "Improving Round-Trip Time Estimates in Reliable Transport Protocols," *Computer Communication Review*, Vol. 17, No. 5, pp. 2–7, Aug. 1987.

15. M. Garetto, R. Lo Cigno, M. Meo, M. Ajmone Marsan, "Modeling Short-Lived TCP Connections with Open Multiclass Queuing Networks – Extended Version," Politecnico di Torino ech. Rep. DE/RLC/2001-4, Politecnico di Torino, June 2001. Available at http://www.tlc-networks.polito.it/locigno/papers/de-rlc-01-4.ps

TCP over High Speed Variable Capacity Links: A Simulation Study for Bandwidth Allocation

Henrik Abrahamsson[1], Olof Hagsand[2], and Ian Marsh[1]

[1] SICS AB, Kista S-164 29, Sweden,
{henrik,ianm}@sics.se
[2] Dynarc AB, Kista S-164 32, Sweden,
hagsand@dynarc.se

Abstract. New optical network technologies provide opportunities for fast, controllable bandwidth management. These technologies can now explicitly provide resources to data paths, creating demand driven bandwidth reservation across networks where an applications bandwidth needs can be meet almost *exactly*. Dynamic synchronous Transfer Mode (DTM) is a gigabit network technology that provides channels with dynamically adjustable capacity. TCP is a reliable end-to-end transport protocol that adapts its rate to the available capacity. Both TCP and the DTM bandwidth can react to changes in the network load, creating a complex system with inter-dependent feedback mechanisms. The contribution of this work is an assessment of a bandwidth allocation scheme for TCP flows on variable capacity technologies. We have created a simulation environment using ns-2 and our results indicate that the allocation of bandwidth maximises TCP throughput for most flows, thus saving valuable capacity when compared to a scheme such as link over-provisioning. We highlight one situation where the allocation scheme might have some deficiencies against the static reservation of resources, and describe its causes. This type of situation warrants further investigation to understand how the algorithm can be modified to achieve performance similar to that of the fixed bandwidth case.

Keywords: TCP, DTM, rate control, rate adaption

1 Introduction

Reliable transfer of data across the Internet has become an important need. TCP [Pos81] is the predominant protocol for data transfer on the Internet as it offers a reliable end-to-end byte stream transport service. Emerging optical networking technologies provide fast, cheap and variable capacity bandwidth links to be setup in milliseconds allowing data-driven virtual circuits to be created when needed. One example of an application that could use such a service is the backup of critical data.

Exact allocation of bandwidth to TCP flows would alleviate complex traffic engineering problems such as provisioning and dimensioning. Allocating bandwidth to TCP is a complex problem; the TCP congestion control mechanism

G. Carle and M. Zitterbart (Eds.): PfHSN 2002, LNCS 2334, pp. 117–129, 2002.

plus network dynamics can make exact allocation for TCP data flows difficult. Te contribution of this paper is the performance evaluation of an estimation algorithm, which measures the rate of TCP flows and allocates capacity on a DTM network.

Dynamic Synchronous Transfer Mode [GHP92] [BHL+96] is a gigabit ring based networking technology that can dynamically adjust its bandwidth. DTM offers a channel abstraction, where a channel consists of a number of slots. The number of slots allocated to a channel determines its bandwidth. The slots can be allocated statically by pre-configured parameters, or dynamically adjusted to the needs of an application. In DTM it is possible to allocate a channel to a specific TCP connection, or to multiplex several TCP connections over the same channel. We mostly investigated cases where each TCP connection is assigned to a separate channel, but show one case in which two TCP connections compete for a single channel. The DTM link capacity is only allocated in the forward direction in this study, we have not performed any allocation for TCP acknowledgements.

TCP uses an end-to-end congestion control mechanism to find the optimal bandwidth at which to send data. In order to get good throughput with TCP operating over a technology such as DTM, it is important to understand the dynamic behaviour of the two schemes, especially when evaluating a bandwidth allocation strategy. TCP is capable of adjusting its *rate* whilst DTM is capable of changing its *capacity*. In dynamically interacting systems, it is possible to create unwanted oscillations resulting in under allocation or over allocation of bandwidth to TCP flows. In order to evaluate the performance of the DTM bandwidth allocator, we have implemented the algorithm in the network simulator ns-2. We have performed a number of simulations that include single and multiple TCP flows, links with varying delay characteristics, different buffer sizes, plus TCP Reno and Tahoe variants.

Section 2 outlines DTM and our estimation algorithm, simulation experiments are given in Section 3, related work follows in Section 4, and finally conclusions and a discussion are given in Section 5.

2   Dynamic Synchronous Transfer Mode

DTM uses a TDM scheme where time slots are divided into control and data slots. The control slots are statically allocated to a node and are used for signalling. Every node has at least one control slot allocated that corresponds to 512 kbps of signalling capacity. The data slots are used for data transmission and each slot is always owned by a node. A node is only allowed to send in slots that it owns. The ownership of the slots is controlled by a distributed algorithm, where the nodes can request slots from other nodes. The algorithms for slot distribution between the nodes affect the network performance. Each slot contains 64 bits and the slots are grouped in 125 microsecond long cycles. The bit rate is determined by the number of slots in a cycle, so one slot corresponds to a bit rate of 512 kbps. By allocating a different numbers of slots, the transmission rate for a channel can be changed in steps of 512 kbps.

2.1   TCP Rate Estimation and DTM Capacity Allocation

TCP's rate is simply estimated as the number of *incoming* bytes per second. The algorithm which is presented next calculates the rate by dividing the number of bytes by the time elapsed. The rate of each flow is calculated ten times per second, i.e. every 100 ms. This value has been chosen as a compromise between good measurement granularity and processing overhead. DTM technology however, has the ability to sample flows up to gigabit speeds, i.e. at sampling rates higher than 100 ms. Actual slot allocation or changes are done only *once* every second, this is slightly coarser due to the overhead of nodes potentially having to negotiate slots.

We now describe the TCP bandwidth estimator. Figure 1 shows the algorithm used to estimate the rate of a given flow. As stated, every 100 ms the estimator measures the rate **new** in bits per second and compares it with the previous value, **current**. A delta of the difference is reduced by DTM_SHIFT in the algorithm. Note this delta is simply shifted, keeping the complexity of the calculation to a minimum. In this case it is three, so the current value is changed by one eighth towards the recently measured flow value, as shown in the first half of the algorithm. This shift effectively determines how aggressively TCP's rate can be tracked. This default value has been chosen experimentally, as DTM is a deployed technology. The technical report version of this paper shows the affect

```
dtm_calc_bw ( new ) {
DTM_SHIFT = 3
MARGIN = 0.75
CORRIDOR = 2

/* first half - Move last estimate closer */
diff = new - current
if ( diff < 0 ) {
    diff = (-diff) ≫ DTM_SHIFT
    current = current - diff // Decreasing
} else {
    diff = diff ≫ DTM_SHIFT
    current = current + diff // Increasing
}
curr_slot = current / slot_bw

/* Second half - Last estimate within bounds ? */
if ( curr_slot > upper_bound ) || (curr_slot < lower_bound ) {
    dynBw = curr_slot + MARGIN + ( CORRIDOR / 2 )
    /* only change bw once per sec */
    change_link_bw (dynBw)
    }
}
```

Fig. 1. Algorithm for bandwidth estimation

of using other values [AM01].Finally the units are changed from bits per second to slots per second by dividing the rate by the channel bandwidth and assigning this value to the variable `curr_slot`.

The second half of the algorithm determines whether it is necessary to change the slot allocations. The current slot value is compared to upper and lower bounds before making any changes. An offset, 0.75 of a slot, `MARGIN` equivalent to 394 kbits, is added to the TCP throughput estimate so the DTM allocation will be a little over the estimated rate. Figure 2 shows two plots using the topology shown in Figure 3, the leftmost plot is the actual measured bandwidth of a single TCP flow. The right plot shows the effect of adding `MARGIN` and measuring the rate in slots. If the allocation was based purely on this estimate it would under allocate bandwidth, causing TCP reduce its window because of congestion on the link. The rightmost graph is coarser due to the second granularity of the bandwidth changes. The plots illustrate how the estimation can be used to give TCP the bandwidth it needs and hence maximise throughput. One can also see in this figure that estimation starts after 100 ms but a change is not applied to the offered bandwidth before the first second. Note also the y-axis in Figure 2b) is in slots per second and not bits per second as in the left figure. Additionally,

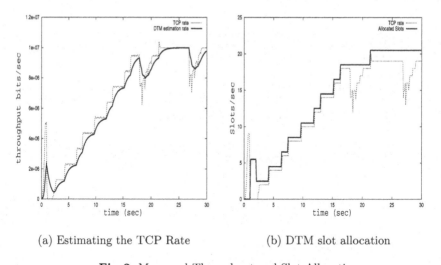

(a) Estimating the TCP Rate   (b) DTM slot allocation

Fig. 2. Measured Throughput and Slot Allocation

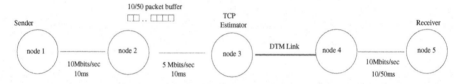

Fig. 3. Simulation topology

a CORRIDOR is an amount the estimate is allowed to vary before slots are added or decreased for a channel. This is not visible in the plots but will be illustrated later. The purpose is to avoid small fluctuations causing unnecessary costly slot allocation changes. As mentioned, slot changes can be time consuming due to the distributed nature of DTM [AMMS98].

3   Simulation Tests

This section presents simulation results that show how the DTM estimation algorithm adapts the offered bandwidth to TCP flows. Figure 3 shows the topology we used for the following simulations. The 5 Mbits per second link between nodes two and three is the bottleneck link. The link between nodes three and four is the DTM link with dynamically allocated bandwidth. Initially the DTM link is set to 10 Mbits per second. This value was chosen simply for convenience, since simulating a 622 Mbits per second link with large bandwidth flows is not feasible in a packet level simulator like ns-2. The other two links also have a capacity of 10 Mbits per second. A bulk transfer TCP Reno flow was setup between nodes one and five and the throughput measured at node three, in order to allocate bandwidth on the outgoing DTM link. In this first simulation the queue length in node 2 was set to 50 packets, figure 4 shows the result. In congestion avoidance the TCP flow increases the congestion window by the maximum segment size bytes each RTT seconds. However, the increase is not made each RTT. Instead TCP will increase $MSS/congestion$ window bytes each time an ACK is received. This means that after RTT seconds, the congestion window was increased by MSS bytes. This continues until the TCP flow has filled the buffer space at the bottleneck link, resulting in a packet drop. TCP Reno, using fast retransmit and fast recovery, then reduces the congestion window by half and continues with congestion avoidance. The congestion window, therefore, follows a sawtooth curve. If enough buffer space is available at the bottleneck link, the rate of the TCP flow, perceived after the second link, is not affected when the congestion window is reduced. This mechanism and result can be seen in left and middle plots of Figure 4. The rightmost plot shows the dynamically allocated bandwidth on the DTM link. It can be seen that TCP actually manages to get about one Megabit per second more on the DTM link due to the extra capacity allocated to the flow through the addition of MARGIN. It should be stated in a real deployment of TCP over DTM that this value is settable by network operators. Its affect can be tested in simulation environments such as this if necessary.

Figure 5 shows the results when the queue size at the bottleneck link is limited to ten packets. This could be the case if a static allocation over the DTM network has been setup. Now the rate of the TCP flow changes with the congestion window, but the changes are too small to affect the dynamic allocation of bandwidth. This is due to the corridor mentioned earlier to avoid small changes from incurring changes in the slot allocation scheme. Figure 6 shows the case in which the simulation with a small queue size and a 50 ms link delay has been repeated using TCP Tahoe instead of TCP Reno. TCP Tahoe

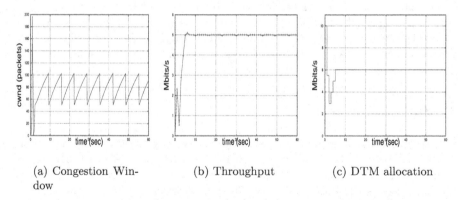

(a) Congestion Window

(b) Throughput

(c) DTM allocation

Fig. 4. Dynamically allocated bandwidth on a DTM link (50 packet queue)

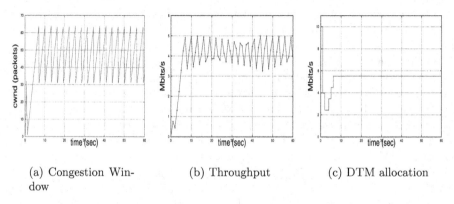

(a) Congestion Window

(b) Throughput

(c) DTM allocation

Fig. 5. Dynamically allocated bandwidth on a DTM link (10 packet queue)

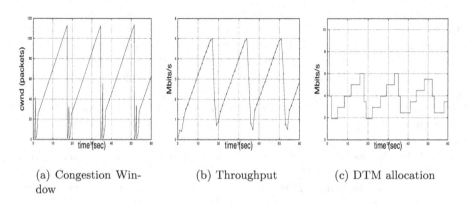

(a) Congestion Window

(b) Throughput

(c) DTM allocation

Fig. 6. Tahoe TCP on a DTM link (10 packet queue)

only relies on the retransmission timer and does not use fast retransmit. When a packet is dropped, the congestion window is set to one and slow-start is invoked. We can see that the allocation on the DTM link closely follows the sharp saw tooth behaviour of TCP Tahoe Figure 6c).

3.1 Two Flows per Channel and Small Router Buffers

So far, we have shown cases where the dynamic allocation of bandwidth has allowed TCP to maximise its throughput. We illustrate one case next when the algorithm has weaknesses to allocate sufficient bandwidth to two TCP Reno flows. In this scenario the fixed link case performs better. Figure 7 shows the topology that we used. It differs from previous simulations in that the flows have their own input buffer at node two but share a common output buffer in the same node. This buffer is also served ten times faster than in previous cases by the fact that the link feeding the DTM network was set to 100 Mbits per second. In this case the queue length of a DTM link, node three, was limited to ten packets.

Figure 8 shows the results when the link capacity between nodes 3 and 4 was fixed at 10 Mbits per second. We can see that both flows manage to reach their 5 Mbits throughput, effectively sharing equally one DTM channel. If we now turn our attention to the same simulation but replace the static link between nodes three and four with a variable one the results are quite different. Figure 9 shows the dynamically allocated bandwidth on the DTM link. Neither of the flows manage to reach 5 Mbits per second on their output links. In this case packets are being dropped in the output buffer of node three. This can be seen in the congestion windows of the two flows, they never manage to maintain the size of the static case, about 100 segments. The problem in this case is the estimation algorithm should not *decrease* the estimation *if* packets are being dropped. The algorithm is symmetric, it increases or decreases depending on the measured rate. Additionally the effect of the short queue does not help, there is not sufficient pressure with a small queue to keep the rate up, with a larger buffer there is more pressure due to accumulated packets. Interestingly, the algorithm actually

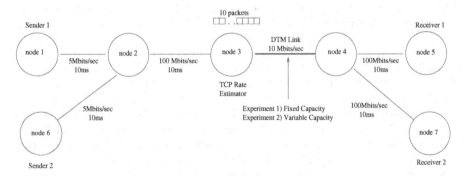

Fig. 7. Simulation topology with two flows

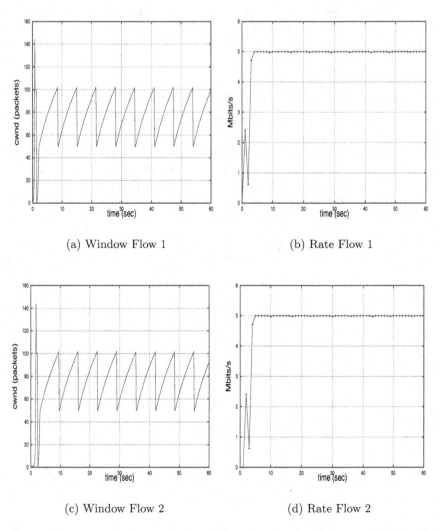

(a) Window Flow 1   (b) Rate Flow 1

(c) Window Flow 2   (d) Rate Flow 2

Fig. 8. Experiment 1 static link: The senders do not drop packets at the ingress node and achieve their constrained link throughput 5Mbits per second.

correctly allocates for the observed throughput, however does not maximise the TCP throughput.

Some researchers have put forward TCP variants which are capable of esti-mating the bandwidth such as TCP Westwood [MG98], which do not solely rely on packet loss for congestion. It is not clear whether TCP variants such as this will improve on the situation above without substantial simulations. However other TCP variants would be worthy of investigation. The important point is note is it is important to detect loss early and this can be done by monitoring

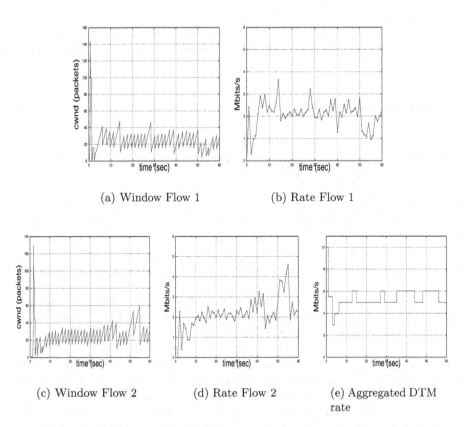

(a) Window Flow 1   (b) Rate Flow 1

(c) Window Flow 2   (d) Rate Flow 2   (e) Aggregated DTM
                                                        rate

Fig. 9. Experiment 2 Dynamic Link: Allocated bandwidth on DTM link. Congestion window and measured rate for two TCP Reno flows with limited queue size at the DTM link.

queues for the local node or even via mechanisms such as RED or ECN for upstream nodes.

Our conclusion is that in the face of loss at a node the estimation algorithm should not decrease to allow TCP to recover. So a special case for loss could be introduced, *during loss* the rate could be estimated but no action is taken to adjust the bandwidth. One other solution would be to adjust the rate more slowly when allocating *less* bandwidth. This is trivial in the present scheme, the DTM_SHIFT variable can be split into DTM_SHIFT_INC and DTM_SHIFT_DEC where a decrease in bandwidth takes places at a slower rate. This might have adverse affects on a normal behaving system so, once again would need to be validated with further simulations. We note that [Kri01] also performed comparisons with the fixed link case and show in one experiment that the dynamically allocated link was not able to achieve the throughput of a fixed link case. With a certain selection of parameters it was only possible to allocate slightly over half of the fixed link case. We conclude that it is not always possible to perform as the fixed link case, however comparisons, wherever possible should be performed. The savings of allocating just the required bandwidth, however can be considerable. This was the methodology employed during our investigation.

4   Related Work

Work on estimating and maximising TCP throughput for variable capacity links is relatively scarce. However comprehensive studies have been done related to the performance of TCP on ATM networks [Bon98] [MG95] [CLN96]. The main conclusions of the works are similar, the traffic classes of ATM are poorly suited to the bursty needs of TCP, due to the traffic contracts needed by ATM classes. The conclusion of [Bon98] is that the complexity of choosing traffic parameters for ABR is not in proportion to the benefits of carrying TCP/IP traffic. The CBR class is too simple for TCP, as only the peak rate is specified. Most of the DTM research in this area focuses on the distributed slot allocation for example [AMMS98].

Clark and Fang propose a framework for allocating bandwidth to different users during congestion [CF98]. The focus of the work is TCP bulk-data transfers. The authors attempt to keep TCP flows in congestion avoidance in the best case, and fast recovery phase in the worst case, by avoiding dropping several packets of the same flow in the same RTT. The conclusions of the given work are similar to those of [Bon98], that TCP connections can have difficulties to fill their alloted bandwidth. The work resembles ours in that they attempt to allocate bandwidth between different flows in a fair manner. It differs from ours in that we assume that the network can change its offered bandwidth and we focus on maximising TCP throughput, rather than trying to maintain a TCP state in the face of adverse network conditions. In addition, we allocate bandwidth to flows not only when the network is congested but also in normal situations as well.

Sterbenz and Krishnan investigate TCP over Load-Reactive Links in a ICNP publication [KS01] and a technical memorandum [Kri01]. They use a hysteresis control mechanism for capacity allocation. Buffer levels are monitored (as in [Lun98]) and if the occupancy is greater than a threshold the capacity is increased and vice versa. This approach is not the same as ours, we measure the rate of incoming TCP flows at the router before the DTM link rather than the buffer level in the router at the outgoing DTM link. Their scheme is dependent on keeping buffers occupied all the time, otherwise the link capacity will fall and hence the throughput. A single TCP flow is simulated and the authors state that the control parameters should be carefully chosen. Poor parameter choice can have the opposite effect, resulting in TCP not being able to operate, as stated previously. The work resembles ours in that a method is presented to react to network load and allocate bandwidth for TCP accordingly. It differs in that we measure the throughput of individual flows and allocate bandwidth from this measurements, where they use the buffer length as a measure of the load. Our system is less scalable, but more accurate, as we can ascertain exactly the bandwidth of the incoming TCP connections. Also there is no need to keep buffers occupied to allocate bandwidth for TCP connections. Also there is no potential interaction between several dynamic allocation schemes running on the same node.

Lundqvist evaluates different algorithms for bandwidth allocation for DTM channels transporting IP traffic [Lun98]. The algorithms were assessed with respect to throughput, delay and bandwidth changes per second. TCP rate adjustment is done by placing the incoming packets into a buffer and adding and removing slots if the level of the buffer exceeds continuously maintained threshold values. He concludes that adaptive strategies are recommended for TCP, however too frequent changes can be undesirable in a DTM network due to the processing cost. The main conclusion from this work is that the choice of algorithm can play a significant role in the performance. This work is similar to ours in that the goal is a slot allocation for TCP traffic over DTM. We also agree it is important to keep the computational complexity low and DTM bandwidth changes as infrequent as possible. It differs from ours in that we measure the rate of each TCP flow, whilst he looks at the outgoing buffer length as a sign to increase or decrease the number of slots. We look more into network scenarios such as different link delays, buffer lengths and use two different TCP types, TCP Reno and Tahoe.

5   Conclusions

We have analysed a complex problem, allocating bandwidth to a protocol that can adapt its rate. The benefits of guaranteeing throughput for an application using TCP can be very beneficial, in particular the cost savings when paying per unit of transmission. The goal was to investigate the behaviour of our bandwidth estimation scheme, its affect on TCP and on a network that can vary its capacity, in this case DTM. Our work however is not only limited to DTM technology, we can draw the same conclusions about TCP performance on any high speed network technology that offers variable capacity.

We have written a simulation environment using ns-2, and found that in almost all cases, TCP could be allocated a share of the channel identical to its measured throughput on a fixed network. We identified one scenario in which the algorithm could be improved, when packets are dropped at a router with a small buffer. In this situation the estimation algorithm should not reduce the offered bandwidth further, resulting in less offered bandwidth and further packet loss. Instead it should allow TCP to find the new capacity available in the network. The combination of the small buffer size plus high speed input link aggravates this observed deficiency.

There are some other open issues, the system as described relies on measuring individual TCP flows, therefore methods that encapsulation such as MPLS would hinder us from measuring single flows. An alternative is monitor and measure aggregate flows, however this was not the focus of our work. DTM is a fast MAN technology and can monitor flows at gigabit speeds, however if a bandwidth allocation scheme would be used in a non-DTM environment, especially in a backbone, some consideration would be needed for the sampling and measuring rate one can achieve with thousands of TCP flows. Aggregation of flows in this context would be a viable alternative.

We have only considered bulk data transfers, as the scheme measures flow bandwidths, it is not feasible to allocate bandwidth for all TCP flows, particularly http transfers as most data is transferred during the slow-start phase of a TCP connection, e.g. banners, buttons etc. Another issue is capacity provisioning for ACKs, we assumed the return path for acknowledgements is not constrained. In our experiments we had a return channel of 512 kbits per second which was more than adequate to support the forward data rates we were using, a maximum of 100Mbits per second. Further investigation is needed to state where problems could arise as well potential solutions.

We have not considered well known sceneries such as satellite links with large bandwidth-delay products or more interestingly, where the control loops are sensitive to delay. We believe some benefit would be gained by looking at this problem (and others with time sensitive mechanisms) from a control theory perspective rather than the traditional networking approach, Westwood referenced earlier, takes exactly this approach.

In a simulation environment the parameter space is large. Due to space limitations we have only discussed a key subset of possible buffer sizes, link bandwidths, link delays and TCP variants. Parameters that are worthy of further investigation include sampling times and estimation thresholds. Further results, plus validation tests for using ns-2 in these kind of simulations, can be found in the technical report [AM01].

Acknowledgements

We would sincerely like to acknowledge the Computer and Network Architecture lab at SICS, and the Laboratory of Communication Networks at KTH, Royal Institute of Technology, Stockholm.

References

[AM01]   Henrik Abrahamsson and Ian Marsh. Dtmsim – dtm channel simulation in ns. Technical Report T2001:10, SICS – Swedish Institute of Computer Science, November 2001.

[AMMS98]   Csaba Antal, József Molnár, Sándor Molnár, and Gabor Szabó. Performance study of distributed channel allocation techniques for a fast circuit switched network. *Computer Communications*, 21(17):1597–1609, November 1998.

[BHL+96]   Christer Bohm, Markus Hidell, Per Lindgren, Lars Ramfelt, and Peter Sjödin. Fast circuit switching for the next generation of high performance networks. *IEEE Journal on Selected Areas in Communications*, 14(2):298–305, February 1996.

[Bon98]   O. Bonaventure. *Integration of ATM under TCP/IP to provide services with minimum guaranteed bandwidth.* PhD thesis, University of Liege, 1998.

[CF98]   D. Clark and W. Fang. Explicit allocation of best-effort packet delivery service. *IEEE/ACM Transactions on Networking*, 6(4), August 1998.

[CLN96]   E. Chan, V. Lee, and J. Ng. the performance of bandwidth allocation strategies for interconnecting atm and connectionless networks, 1996.

[GHP92]   L. Gauffin, L. H. Hakansson, and B. Pehrson. Multi-gigabit networking based on DTM. *Computer Networks and ISDN Systems*, 24(2):119–130, April 1992.

[Kri01]   Rajesh Krishnan. Tcp over load-reactive links. Technical Memorandum 1281, BBN, February 2001.

[KS01]   Rajesh Krishnan and James Sterbenz. Tcp over load-reactive links. In *International Conference on Network Protocols (ICNP)*, 2001.

[Lun98]   Henrik Lundqvist. Performance evaluation for IP over DTM. Master's thesis, Linköping University, Linköping, 1998.

[MG95]   Kjersti Moldeklev and Per Gunningberg. How a large ATM MTU causes deadlocks in TCP data transfers. *IEEE/ACM Transactions on Networking*, 3(4):409–422, 1995.

[MG98]   Saverio Mascolo and Mario Gerla. TCP congestion avoidance using explicit buffer notification. Technical report, University of California, Los Angeles, California, February 1998.

[Pos81]   J. Postel. Transmission control protocol. Request for Comments 793, Internet Engineering Task Force, September 1981.

TCP Westwood and Easy RED to Improve Fairness in High-Speed Networks

Luigi Alfredo Grieco[1] and Saverio Mascolo[2]

[1] Dipartimento d'Ingegneria dell'Innovazione, Università di Lecce, Italy
`alfredo.grieco@unile.it`
[2] Dipartimento di Elettrotecnica ed Elettronica, Politecnico di Bari, Italy
`mascolo@poliba.it`

Abstract. TCP Westwood (TCPW) is a sender-side only modification of TCP Reno congestion control, which exploits end-to-end bandwidth estimation to properly set the values of *slow-start threshold* and *congestion window* after a congestion episode. This paper aims at showing via both mathematical modeling and extensive simulations that TCPW significantly improves fair sharing of high-speed networks capacity and that TCPW is friendly to TCP Reno. Moreover, we propose EASY RED, which is a simple Active Queue management (AQM) scheme that improves fair sharing of network capacity especially over high-speed networks. Simulation results show that TCP Westwood provides a remarkable Jain's fairness index increment up to 200% with respect to TCP Reno and confirm that TCPW is friendly to TCP Reno. Finally, simulations show that Easy RED improves fairness of Reno connections more than RED, whereas the improvement in the case of Westwood connections is much smaller since Westwood already exhibits a fairer behavior by itself.

1 Introduction

Packet switching networks require sophisticated mechanism of flow and congestion control in order to share resources and avoid congestion phenomena. Congestion control functions were introduced into the TCP in 1988 and have been of crucial importance in preventing congestion collapse [1],[2],[9]. However, while end-to-end TCP congestion control [4],[5] can ensure that network capacity is not exceeded, it cannot insure fair sharing of that capacity [1]. In this paper we investigate via both mathematical analysis and computer simulations the issue of fairness in high-speed networks when Westwood TCP is implemented at the sender side. Moreover we propose a simpler version of RED, called EASY RED and we investigate how it interacts with Reno and Westwood TCP.

TCP Westwood (TCPW) performs an end-to-end estimate of the bandwidth available along a TCP connection to adaptively set the control windows after congestion [3]. The rationale of TCPW is simple: in contrast with TCP Reno, which implements a multiplicative decrease algorithm after congestion, TCPW sets a slow

G. Carle and M. Zitterbart (Eds.): PfHSN 2002, LNCS 2334, pp. 130–146, 2002.

start threshold and a congestion window which are consistent with the effective bandwidth used at the time congestion is experienced.

In this paper, TCPW employs a bandwidth estimation algorithm that is slightly different from the one used in [3] in order to avoid bandwidth overestimates due ACK compression [6],[11].

EASY RED is a simpler variant of RED that does not average the queue length but relates the drop probability to the instantaneous queue level. In fact, the purpose of early discard is to signal congestion to the sender as soon as possible. In contrast averaging the queue introduces delay, which is harmful for congestion control purposes. EASY RED has only two parameters to be set: (1) the minimum threshold (*min_th*) and (2) the constant drop probability *pdrop* when the instantaneous queue length is greater or equal to *min_th*.

A main contribution of this paper is a mathematical model that proves stability, fairness and friendliness of TCP Westwood with respect to Reno. In particular, the model shows that the mean throughput of TCP Westwood is function of the available bandwidth and is less sensitive to round trip time than Reno throughput, that is, Westwood improves fair sharing of network capacity among flows with different RTTs. Moreover, the model highlights that the throughput of TCPW depends on the inverse of the square root of the drop probability just like the throughput of Reno [18],[25], that is, TCPW is friendly to TCP Reno.

Simulation results using Westwood show a remarkable increment of the Jain fairness index up to 200% with respect to Reno over a 100Mbps wired network. Also they confirm the theoretical model by showing that TCPW is completely friendly to Reno.

Performance improvements are also shown when AQM mechanisms are used. Simulations show that EASY RED improves fairness of Reno connections more than RED, whereas the improvement in the case of Westwood connections is much smaller since Westwood already exhibits a fairer behavior by itself.

The paper is organized as follows: in Section 2 a mathematical model of TCP Westwood is developed; in Section 3, Active Queue Management algorithms are described and Easy RED is proposed; in Section 4, simulation results with many Reno or Westwood TCP connections having different RTTs and sharing a FIFO bottleneck queue implementing RED, Gentle RED, EASY RED or no AQM policy are reported. Finally, Section 5 draws the conclusions.

2   TCP Westwood

A detailed description of TCP Westwood (TCPW) is reported in [3]. In this section, we briefly resume TCPW and we introduce a new mechanism to estimate the available bandwidth. Later we develop a mathematical model of Westwood and analyze fairness and friendliness of Westwood in comparison with Reno by using their respective throughput equation models.

2.1   A Description of TCP Westwood

A TCP connection is characterized by the following variables:

- *cwnd*: Congestion Window
- *ssthresh*: Slow Start Threshold
- *RTT*: Round Trip Time of the connection
- RTT_{min}: Minimum Round Trip Time measured by the sender
- *seg_size*: Size of the delivered segments

The main idea of TCP Westwood is to perform an end-to-end estimate of the bandwidth *B* available along a TCP connection by measuring and low-pass filtering the rate of returning ACKs. For available bandwidth we mean the measurement of the actual rate a connection is achieving during the data transfer. This is a much more easy task than estimating the bandwidth that is available at the beginning of a TCP connection [12],[14],[15],[16]. The bandwidth estimate is then used to properly set the congestion window and the slow-start threshold after a congestion episode as described below:

```
a)   When 3 DUPACKs are received by the sender:
     ssthresh = (B*RTT_min)/seg_size;
     cwnd = ssthresh;

b)   When coarse timeout expires:
     ssthresh = (B*RTT_min)/seg_size;
     cwnd = 1;

c)   When ACKs are successfully received:
     cwnd is increased following the Reno algorithm.
```

As it has been pointed out in [1],[2],[26], the stability of the Internet does not require that flows reduce their sending rate by half in response to a single congestion indication. In particular, the prevention of congestion collapse simply requires that flows use some form of end-to-end congestion control to avoid a high sending rate in the presence of high packet drop rate. In the case of TCPW the sending rate is reduced by taking into account a measurement of the available bandwidth at the time congestion is experienced. Therefore, when in the presence of heavy congestion, this reduction can be even more drastic than a by half reduction and it can be less drastic with light congestion. This feature can clearly improves network stability and utilization in comparison with the *blind* by a half window reduction performed by Reno.

2.2   Robustness of Bandwidth Estimate to ACK Compression

In order to fully exploit the advantages of the AIAD paradigm, it is of crucial importance to obtain a *good estimate* of the bandwidth that is available when congestion is experienced. Due to delays and ACKs compression, the flow of returning ACKs must be low-pass filtered in a proper way [11],[17]. In [3], a sample of available bandwidth $b_k = d_k /(t_k - t_{k-1}) = d_k / \Delta_k$ is computed every time t_k the sender receives an ACK, where the amount d_k of data acknowledged by an ACK is determined by a proper counting procedure that takes into account delayed ACKs, duplicate and cumulative ACKs. Samples b_k are low-pass filtered using the time-varying filter $\hat{b}_k = \dfrac{2\tau_f - \Delta_k}{2\tau_f + \Delta_k}\hat{b}_{k-1} + \Delta_k \dfrac{b_k + b_{k-1}}{2\tau_f + \Delta_k}$, where τ_f is the filter time constant (a typical value is $\tau_f = 0.5$ s). In this paper, we propose a slightly modified version of the filter used in [3] since that filter overestimates the available bandwidth when in the presence of ACK compression [6]. To overcome this problem, we compute bandwidth samples every *RTT*. More precisely, we count all data d_k acknowledged during the last *RTT* and compute the bandwidth sample $b_k = d_k / \Delta_k$, where Δ_k is the last *RTT*. Moreover, in order to comply with the Nyquist sampling theorem when $\Delta_k > \tau_f / 4$, we interpolate and re-sample using $N = int(4 \cdot RTT / \tau_f)$ [1] virtual samples b_k arriving with the interarrival time $\Delta_k = \tau_f / 4$.

In order to test the robustness of the new filter with respect to ACK compression, we simulate a single bottleneck scenario shared by one TCP and one UDP connection via FIFO queuing. The bottleneck link capacity is 1Mbps. In order to provoke ACK compression, 10 TCP Reno connections sending data along the reverse path are considered. Segment size is 1500 Bytes long, queue size is 20 segments and the simulation lasts 1000s. Fig. 1(a) shows the bandwidth estimate obtained using the old and the new filter when the UDP source is turned off. The tick lines marks the available bandwidth that is 1Mbps. Fig. 1(a) shows that the old filter overestimates the bandwidth ten times, whereas the new one nicely tracks the available bandwidth. Fig. 1(b) shows the bandwidth estimate obtained when the UDP sources is active. The tick line marks the bandwidth that is left available by the UDP traffic. Also in this case the new filter tracks the available bandwidth whereas the old one overestimates up to 10 times the available bandwidth.

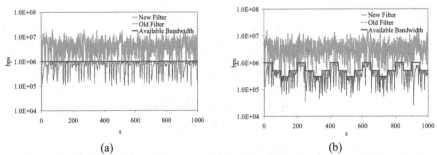

(a)   (b)

Fig. 1. Bandwidth estimates: (a) without UDP traffic; (b) with coexisting UDP traffic

2.3   A Mathematical Model of TCP Westwood

In this section a mathematical model of the *Additive Increase Adaptive Decrease* mechanism introduced by Westwood is developed. To derive the model, we follow

---

[1] *int*($\cdot$) stands for the integer part of ($\cdot$)

arguments similar to the ones developed in the excellent paper by Kelly [18]. For the sake of simplicity, we do not model the behavior after a timeout.

Theorem 1. Consider a TCP flow that is controlled by the Westwood algorithm. Suppose that the drop probability of a segment is p, the bandwidth available for the flow is B, the mean round trip time is RTT and the minimum round trip time is RTT_{min}. By letting r^W be the steady state mean throughput of the flow, it holds:

$$r^W = \frac{B \cdot RTT_{min}}{2 \cdot RTT} + \sqrt{\left(\frac{B \cdot RTT_{min}}{2 \cdot RTT}\right)^2 + \frac{1-p}{p \cdot (RTT)^2}} \; . \tag{1}$$

Proof. The congestion window is updated upon ACK reception. Each time an ACK is received back by the sender the *cwnd* is increased by 1/*cwnd*, whereas after a segment loss the congestion window is set equal to $B \cdot RTT_{min}$ so that the change in *cwnd* is $B \cdot RTT_{min} - cwnd$. Since the segment drop probability is p, it follows that the expected increment of the congestion window *cwnd* per update step is:

$$E[\Delta cwnd] = \frac{1-p}{cwnd} + (B \cdot RTT_{min} - cwnd) \cdot p \; . \tag{2}$$

Since the time between update steps is about $\dfrac{RTT}{cwnd}$, by recalling Eq. (2), the expected change in the rate r per unit time is approximately:

$$\frac{\partial r(t)}{\partial t} = \frac{1-p}{RTT^2} + p \cdot r(t) \cdot \frac{B \cdot RTT_{min}}{RTT} - r^2(t) \cdot p \; . \tag{3}$$

Eq. (3) is a separable variable differential equation. By separating variables, Eq. (3) can be written as:

$$\frac{\partial r(t)}{r^2(t) - r(t) \cdot \dfrac{B \cdot RTT_{min}}{RTT} - \dfrac{1-p}{p \cdot RTT^2}} = -p \cdot \partial t \; . \tag{4}$$

and by integrating each member the following solution can be obtained

$$r(t) = \frac{r_1 - r_2 \cdot C \cdot e^{-p \cdot t \cdot (r_1 - r_2)}}{1 - C \cdot e^{-p \cdot t \cdot (r_1 - r_2)}} \; .$$

Where $r_{1,2} = \dfrac{B \cdot RTT_{min}}{2 \cdot RTT} \pm \sqrt{\left(\dfrac{B \cdot RTT_{min}}{2 \cdot RTT}\right)^2 + \dfrac{1-p}{p \cdot (RTT)^2}}$ are the roots of the equation

$r^2 - r \cdot \dfrac{B \cdot RTT_{min}}{RTT} - \dfrac{1-p}{p \cdot RTT^2} = 0$ and C depends on the initial conditions.

The steady state throughput of the Westwood algorithm is then

$$r^W = \lim_{t \to \infty} r(t) = \frac{B \cdot RTT_{\min}}{2 \cdot RTT} + \sqrt{\left(\frac{B \cdot RTT_{\min}}{2 \cdot RTT}\right)^2 + \frac{1-p}{p \cdot (RTT)^2}}$$ ∎

It is easy to show the following corollary.

Corollary 1. The Westwood control algorithm is stable, that is

$$r^W \le B .$$ (5)

Proof. From Eq. (1) we can argue that r^W can never be greater than B. In fact, by contradiction, let us assume that $r^W > B$. This assumption would lead to congestion collapse so that the drop probability p would increase up to 1. Thus, from Eq. (1) it would result $r^W = B \frac{RTT_{\min}}{RTT}$. Since under congestion $RTT_{\min} < RTT$, it would result $r^W < B$, which would contradict the assumption. Therefore, we can conclude that $r^W \le B$.   ∎

Now, by noting that the end-to-end bandwidth estimation algorithm described above provides a value that is well approximated by *cwnd/RTT*, it is possible to mathematically derive the throughput of Westwood when the bandwidth estimation algorithm described in this paper is employed.

Theorem 2. The steady state throughput of Westwood using the bandwidth estimate *B=cwnd/RTT* is

$$r^{West} = \frac{1}{\sqrt{RTT \cdot T_q}} \cdot \sqrt{\frac{(1-p)}{p}} .$$ (6)

Where $T_q = RTT - RTT_{min}$ is the mean queuing time experienced by the segments of the connection.

Proof. By assuming the following estimate of the available bandwidth

$$B = cwnd / RTT = r(t)$$ (7)

and by substituting Eq. (7) into Eq. (3), the following differential equation is obtained:

$$\frac{\partial r(t)}{\partial t} = \frac{1-p}{RTT^2} + p \cdot r^2(t) \cdot \frac{RTT_{\min}}{RTT} - r^2(t) \cdot p .$$ (8)

By separating variables, Eq. (8) can be written as:

$$\frac{\partial r(t)}{r^2(t) - \frac{1-p}{p \cdot T_q \cdot RTT}} = -p \cdot \frac{T_q}{RTT} \partial t \,. \tag{9}$$

and integrated as:

$$r(t) = \sqrt{\frac{1-p}{p \cdot RTT \cdot T_q}} \cdot \frac{1 + C \cdot e^{-2 \cdot p \cdot t \cdot \frac{T_q}{RTT} \sqrt{\frac{1-p}{p \cdot RTT \cdot T_q}}}}{1 - C \cdot e^{-2 \cdot p \cdot t \cdot \frac{T_q}{RTT} \sqrt{\frac{1-p}{p \cdot RTT \cdot T_q}}}} \,. \tag{10}$$

Where C depends on the initial conditions. The steady state throughput (6) is then obtained for $t \to \infty$. ∎

2.3   Fairness and Friendliness Evaluation

Kelly derives the following steady state mean throughput of Reno TCP [18]:

$$r_R = \frac{1}{RTT} \cdot \sqrt{\frac{2 \cdot (1-p)}{p}} \,. \tag{11}$$

With reference to friendliness, by comparing (6) and (11) it can be noted that both throughputs of Westwood and Reno depend on $1/\sqrt{p}$, that is Westwood and Reno are friendly to each other. Moreover, Eq. (6) shows that flows with different $RTTs$ and going through the same bottleneck, experience the same mean queuing time T_q. Therefore, the throughput of Westwood depends on round trip time as $1/\sqrt{RTT}$ whereas throughput of Reno as $1/RTT$, that is, Westwood increases fair sharing of network capacity between flows with different RTTs.

3   AQM Policies and Easy RED

The idea behind Active Queue Management (AQM) is to discard a packet before queue overflow in according to a drop probability function. The rationale is that, by discarding a packet before queue overflow, a TCP sender can detect congestion earlier and react earlier.

The most know example of AQM mechanism is RED, which uses a drop probability function that increases linearly with the average queue length [19]. RED needs the tuning of four parameters that are: (1) the minimum queue threshold (min_th); (2) the maximum queue threshold (max_th); (3) the drop probability max_p when the average queue reaches the max_th and (4) the constant value used by the exponential filter to average the queue length. A delicate issue with RED is that it requires fine-tuning of many parameters in order to work properly. Consequently, there is considerable nervousness in the community regarding the deployment of RED [10],[20],[21],[22].

Several complex variants of RED have been proposed in order to obtain algorithms less sensitive to parameter tuning. In [29], stabilized RED (SRED) is proposed, which aims at stabilizing buffer occupation by estimating the number of active connections in order to set the drop probability as a function of the number of the active flows and of the instantaneous queue length. In [28], Flow RED (FRED) is proposed which uses per-active-flow accounting to impose on each flow a loss rate that depends on the flow's buffer use. FRED employs the same drop probability function of RED; furthermore, it maintains minimum and maximum limits on the packets that a flow may have in the queue and uses a more aggressive drop against the flows that violates the maximum bound. In [27] and [32] schemes to auto tune RED parameters are proposed. These schemes essentially increase the max_p parameter when the average queue length exceeds a fixed target and decrease max_p when the average queue length falls below the target level. The Balanced RED algorithm, which tries to regulate the bandwidth assigned to each flow by doing per flow accounting, is proposed in [30]. BRED stores the per flow buffer level and for each incoming packet it computes the drop probability as a function of the buffer level of the flow to which the packet belongs. Finally Dynamic RED [31] proposes to discard packets with a load dependent probability. In particular DRED continuously update the drop probability by employing an integral controller with a gain in cascade. The input of the controller is the difference between the average queue length and the target buffer level whereas the output is the drop probability.

In this section, we introduce a simpler variant of RED that we call EASY RED. We show that EASY RED improves fairness and that it is not sensitive to parameters tuning. EASY RED does not average the queue length but it relates the drop probability to the instantaneous queue level. In fact, the purpose of early discard is to signal congestion to the sender as soon as possible. In contrast to this, the queue average of RED introduces delay, which is harmful for congestion control purposes. In control terms, averaging means the introduction of an extra pole in the control loop [20],[22].

EASY RED has only two parameters to be tuned: (1) the *min_th* and (2) the constant drop probability when the instantaneous queue length is greater or equal to *min_th*. Fig. 2 shows the dropping profile of EASY RED and RED. EASY RED has a flat dropping probability that is function of the *instantaneous queue length*, whereas RED has a linearly increasing drop probability that jumps to one when the *average queue length* reaches the *max_th* [23]. The gentle variant of RED eliminates the jump to one using another linear piece of curve [24].

4   Performance Evaluation

In this section, we test TCPW using the ns-2 simulator [7] and we validate results obtained in Sec. 2, which are: (1) TCP Westwood improves fairness; (2) TCP Westwood is friendly to Reno. Moreover we test the behavior of EASY RED.

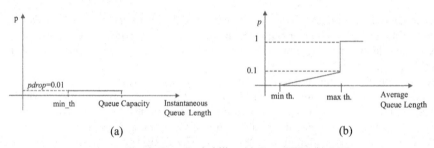

Fig. 2. Drop Probability. (a) Easy RED; (b) RED

In order to evaluate the performance of TCP Westwood we consider N greedy connections sharing a FIFO bottleneck, with N=5,10,40,70,100 and RTT ranging uniformly from (250/N)ms to 250ms. The sources transmit data during a period of 30 seconds. The segment size is 1500 Bytes long. The bottleneck link bandwidth is set equal to 10Mbps or 100Mbps and the bottleneck queue capacity is set equal to the link capacity times the maximum round trip propagation time, that is, the bottleneck queue size is set equal to 200 and 2000 segments, respectively. Note that these settings allow a number of segments proportional to the number of flows be accommodated in the bottleneck queue so avoiding the *many flows* effect [13]. To provide a single numerical measure reflecting the fair share distribution across the various connections we use the Jain Fairness Index defined as:

$$Fairness\ Index = \frac{(\sum_{i=1}^{N} b_i)^2}{N \sum_{i=1}^{N} b_i^2}$$

where b_i is the throughput and N is the number of connections [8].

4.1   Fairness of TCP Westwood

In this section, we compare the fairness of TCPW versus the fairness of TCP Reno without using AQM policies. Fig. 3 (a) shows the Jain fairness index as a function of the number of connections when the bottleneck capacity is 10Mbps and Fig. 3 (b) when the bottleneck capacity is 100Mbps. Fig. 3 shows that TCPW improves fairness up to 200% when bottleneck capacity is 100Mbps and up to 60% when bottleneck capacity is 10Mbps.

Figs. 4(a) and (b) show the corresponding mean throughputs computed as the sum of all the throughputs of the N TCP sources sharing the bottleneck divided by N. To give a further insight, Figs 5(a) and (b) show the curves of Bytes_sent vs. time in the case of 40 connection using Reno and in the case of 40 connections using Westwood, respectively. The bottleneck is 100Mbps. Figures clearly show that Reno curves are much more spread than Westwood curves, i.e. TCPW increases fairness.

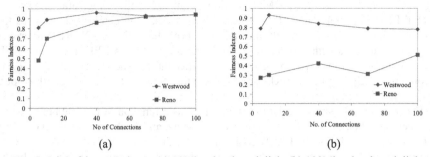

(a)   (b)

Fig. 3. Jain's fairness Indexes: (a) 10Mbps bottleneck link; (b) 100Mbps bottleneck link

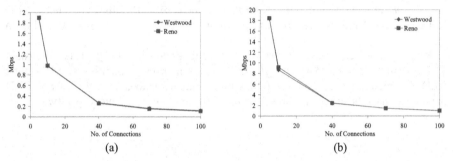

(a)   (b)

Fig. 4. Mean Throughputs: (a) 10Mbps bottleneck link; (b) 100Mbps bottleneck link

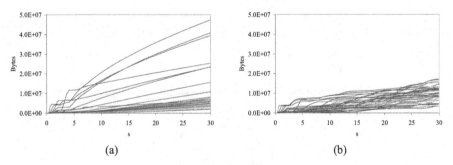

(a)   (b)

Fig. 5. Bytes sent vs. time. (a) 40 Reno connections. (b) 40 Westwood connections

4.2   Interaction with AQM Policies

In this section, we study the effect of AQM policies on the performance of TCP Reno and Westwood. Moreover, we evaluate the performance improvement when Easy RED is employed.

Fig. 6(a) shows the Jain fairness index as a function of the number of Westwood connections when the bottleneck capacity is 10Mbps and Fig. 6(b) when the bottleneck capacity is 100Mbps. Four curves are shown that refer to RED, Gentle

RED, EASY RED and no AQM, i. e. drop from tail, policy. Fig. 6(a) shows that EASY RED does not change the fairness whereas RED and gentle RED reduces the fairness with respect to simple drop of tail. Fig. 6(b) shows that EASY RED improves fairness up to 12% with respect to no AQM policies whereas RED and gentle RED reduces fairness with respect no AQM. Figs. 7(a) and (b) show corresponding mean throughputs: RED and gentle RED reduces the throughput of Westwood with respect to EASY RED and drop tail.

Figs.8(a) and (b) show the Jain fairness index as a function of the number of Reno connections when the bottleneck capacity is 10Mbps and 100Mbps, respectively. Fig. 8 (a) shows that for N<40 EASY RED improves fairness up to 40% whereas RED and gentle RED reduces fairness also respect to no AQM policies for N>10. In the case of Fig. 8(b), EASY RED improves fairness up to 65% with respect to RED policy and up to 165% with respect to drop tail. Figs. 9(a) and (b) show the corresponding mean throughputs. Note that, in the case of 100Mbps bottleneck, RED and gentle RED reduces the Reno throughput with respect to EASY RED and drop tail.

RED parameters have been set as suggested by [23]: filter constant q_weight=0.002, min_th=5, max_th=15, maximum drop probability = 0.1. Gentle RED parameters have been set following recommendations in [24]. EASY RED parameters have been set as follows: $pdrop$=0.01 and min_th=queue_capacity/3.

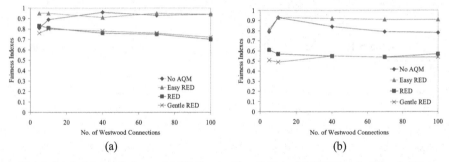

Fig. 6. Jain's fairness Indexes: (a) 10Mbps bottleneck link; (b) 100Mbps bottleneck link

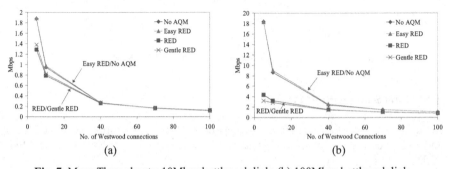

Fig. 7. Mean Throughputs: 10Mbps bottleneck link; (b) 100Mbps bottleneck link

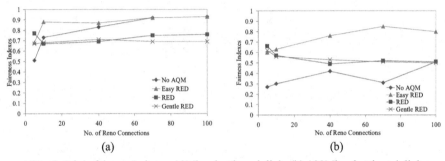

Fig. 8. Jain's fairness Indexes: 10Mbps bottleneck link; (b) 100Mbps bottleneck link

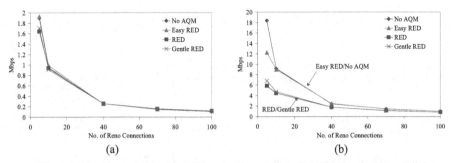

Fig. 9. Mean Throughputs: (a) 10Mbps bottleneck link; (b) 100Mbps bottleneck link

4.3   Sensitivity of Easy RED to Parameters Setting

To investigate the effects that the *pdrop* parameter used in Easy RED has on the fairness and throughput, we vary *prodp* from 0.01 to 0.1 by keeping *min_th=queue_capacity*/3. Figs. 10-13 show that fairness indexes and mean throughputs of Reno and Westwood are not sensitive to *pdrop* over links with capacity of 10Mbps and 100Mbps.

To investigate the effect of the *min_th* parameter we set *min_th=queue_capacity*/*a* and we vary *a* from 2 to 4 by keeping *pdrop*=0.01. Figs. 14-17 show that fairness indexes and mean throughputs of Reno and Westwood are not sensitive to *min_th*.

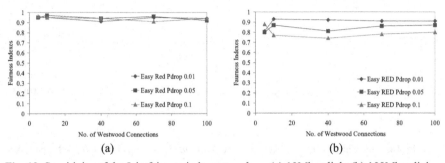

Fig. 10. Sensitivity of the Jain fairness indexes to *pdrop*: (a) 10Mbps link; (b) 100Mbps link

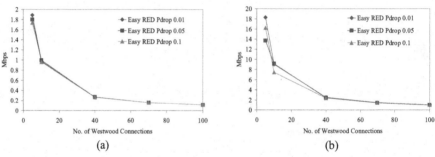

Fig. 11. Sensitivity of mean throughputs to *pdrop*: (a) 10Mbps link; (b) 100Mbps link

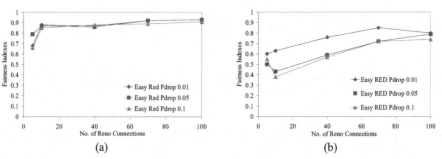

Fig. 12. Sensitivity of the Jain fairness indexes to *pdrop*: (a) 10Mbps link; (b) 100Mbps link

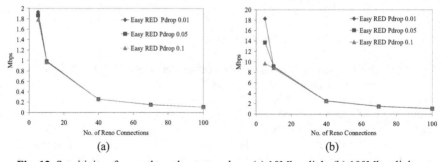

Fig. 13. Sensitivity of mean throughputs to *pdrop*: (a) 10Mbps link; (b) 100Mbps link

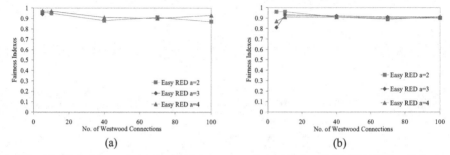

Fig. 14. Jain's fairness Indexes sensitivity to *min_th*: (a) 10Mbps link; (b) 100Mbps link

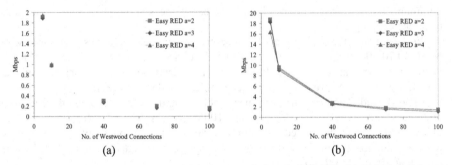

Fig. 15. Mean Throughputs sensitivity to *min_th*. (a) 10Mbps link; (b) 100Mbps link

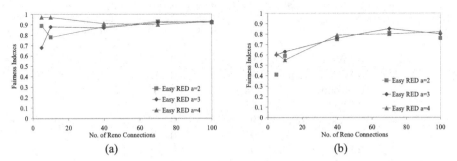

Fig. 16. Jain's fairness Indexes sensitivity to *min_th*: (a) 10Mbps link; (b) 100Mbps link

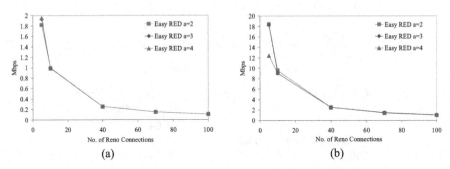

Fig. 17. Mean Throughputs sensitivity to *min_th*: (a) 10Mbps link; (b) 100Mbps link

4.4   Friendliness of TCP Westwood

Friendliness relates to how connections running different TCP flavors affect the performance of one another. In particular, we demand that a newly proposed TCP be *friendly* to the TCP versions already in use in the Internet. That is, the new TCP connections must be able to coexist with connections using existing TCP protocols while providing opportunities for all connections to progress satisfactorily. At a minimum, the new connections should not cause the starvation of the connections using the existing version of TCP.

We simulate the same scenario described in the previous sections with N=10,40,70,100 connections. The Westwood connections are mixed with the Reno connections. In particular, N/2 Reno and N/2 Westwood connections are mixed. Round trip times are spread as in the scenario described in the previous section.

Table 1 and Table 2 show fairness indexes and mean throughputs when the bottleneck link capacity is 10Mbps and 100Mbps, respectively. Results show that indexes obtained in the mixed environments are better than ones obtained with only Reno connections, especially over the 100Mbps high speed link, that is, TCPW is more than friendly to Reno.

Table 1. 10Mbps bottleneck link

Connections	Fairness Index	Mean Throughput (Mbps)
100 West	0.94	0.117
50 West 50 Reno	0.9	0.113
100 Reno	0.94	0.107
70 West	0.93	0.16
35 West 35 Reno	0.9	0.155
70 Reno	0.92	0.148
40 West	0.96	0.266
20 West 20 Reno	0.87	0.261
40 Reno	0.86	0.256
10 West	0.89	0.949
5 West 5 Reno	0.8	0.952
10 Reno	0.7	0.949

Table 2. 100Mbps bottleneck link

Connections	Fairness Index	Mean Throughput (Mbps)
100 West	0.78	1.04
50 West 50 Reno	0.64	1.02
100 Reno	0.51	0.997
70 West	0.79	1.46
35 West 35 Reno	0.66	1.43
70 Reno	0.31	1.43
40 West	0.84	2.39
20 West 20 Reno	0.58	2.38
40 Reno	0.42	2.42
10 West	0.93	8.67
5 West 5 Reno	0.65	8.74
10 Reno	0.3	9.17

5   Conclusions

We have shown, via both mathematical modeling and extensive simulations, that TCP Westwood provides a sensible fairness increment with respect to TCP Reno over high-speed networks. Moreover, we have shown that it is friendly to Reno. We have also introduced a simpler variant of RED, called EASY RED, which improves fairness of Reno connections more than RED, whereas the improvement in the case of Westwood connections is much smaller since Westwood already exhibits a fairer behavior by itself.

References

1. Jacobson, V.: Congestion Avoidance and Control. ACM Computer Communications Review, Vol. 18(4) (1988) 314 - 329
2. Allman, M., Paxson, V., Stevens, W. R.: TCP congestion control. RFC 2581, April 1999
3. Mascolo, S.,Casetti, C., Gerla, M., Sanadidi, M., Wang, R.: TCP Westwood: End-to-End Bandwidth Estimation for Efficient Transport over Wired and Wireless Networks. Proceedings of ACM Mobicom, Rome Italy (2001). To appear in ACM Journal on Wireless Networks (WINET), Special Issue on Wireless Networks with selected papers from MOBICOM2001
4. Clark, D.: The design philosophy of the DARPA Internet protocols. Proceedings of Sigcomm in ACM Computer Communication Review, Vol. 18(4) (1988) 106-114
5. Floyd, S., Fall, K.: Promoting the use of end-to-end congestion control in the Internet. IEEE/ACM Transactions on Networking, Vol. 7(4) (1999) 458-72
6. Mogul, J.C.: Observing TCP dynamics in real networks. Proceedings of Sigcomm in ACM Computer Communication Review, Vol. 22(4) (1992) 305-317
7. ns-2 network simulator (ver 2). LBL, URL: http://www-mash.cs.berkeley.edu/ns
8. Jain, R.: The art of computer systems performance analysis. John Wiley and Sons, (1991)
9. Stevens, W.: TCP/IP illustrated, Addison Wesley, Reading, MA, (1994)
10. Iannaccone, g., May, M, and Diot, C.: Aggregate Traffic Performance with Active Queue Management and Drop from Tail, Computer Communication Review, Vol. 31(3) (2001) 4-13
11. Capone, A., Martignon, F.: Bandwidth Estimates in the TCP Congestion Control Scheme. Tyrrhenian IWDC 2001, Taormina Italy (2001)
12. Hoe, J., C., Improving the Start-up Behavior of a Congestion Control Scheme for TCP. Proceedings of ACM Sigcomm in ACM Computer Communication Review, Vol 26(4) (1996) 270-280
13. Morris, R.: TCP behavior with Many Flows. IEEE International Conference on Network Protocols, Atlanta Georgia (1997) 205-211
14. Keshav, S.: A control-theoretic approach to flow control. Proceedings of Sigcomm in ACM Computer Communication Review, Vol. 21(4) (1991) 3-15
15. Allman M., and Paxson, V.: On Estimating End-to-End Network Path Properties. Proceedings of Sigcomm in ACM Computer Communication Review, (1999) 263-274
16. Lai, K. and Baker, M.: Measuring Link Bandwidths Using a Deterministic Model of Packet Delay. Proceedings of Sigcomm in ACM Computer Communication Review, (2000) 283-294
17. Li, S.Q., and Hwang, C.: Link Capacity Allocation and Network Control by Filtered Input Rate in High speed Networks. IEEE/ACM Transactions on Networking, Vol. 3(1) (1995) 10-25

18. Kelly, F.: Mathematical modeling of the Internet. Proceedings of the Fourth International Congress on Industrial and Applied Mathematics, (1999) 105-116
19. Floyd S., and Jacobson, V.: Random Early Detection gateways for congestion avoidance. IEEE/ACM Transactions on Networking, Vol 1(4) (1997)
20. Hollot, C.V., Misra, V., Towsley, D., and Gong, W,: A control Theoretic Analysis of RED. Proceedings of Infocom (2001)
21. May, M., Bolot, J., Diot, C., Lyles, B.: Reasons not to deploy RED. Seventh International Workshop on Quality of Service IWQoS (1999)
22. Hollot, C.V., Misra, V., Towsley, D., and Gong, W.: On Designing Improved Controllers for AQM Routers Supporting TCP Flows. Proceedings of Infocom (2001)
23. Floyd, S.: RED: Discussions of Setting Parameters, (1997). At http://www.aciri.org/floyd/
24. Floyd, S.: Recommendation on using the "gentle" variant of RED, (2000). At http://www.aciri.org/floyd/
25. Padhye, J., Firoiu, V., Towsley, D., Kurose, J.: Modeling TCP Throughput: A Simple Model and its Empirical Validation. Proceedings of Sigcomm in ACM Computer Communication Review, Vol 28(4) (1998) 303-314
26. Floyd, S., Handley, M., Padhye, J., and Widmer, J.: Equation-Based Congestion Control for Unicast Applications. Proceedings of Sigcomm in ACM Computer Communication Review, Vol. 18 (2000) 43-56
27. Feng, W., Kandlur, D., Saha, D., Shin, K.G.: A Self-Configuring RED Gateway. Proceedings of Infocom (1999)
28. Lin, D., and Morris, R.: Dynamics of Random Early Detection. Proceedings of Sigcomm in ACM Computer Communication Review, Vol. 27(4) (1997) 127-137
29. Ott, T.J., Lakshman, T.V., Wong, L.: SRED: Stabilized RED. Proceedings of Infocom (1999)
30. Anjum, F.M., and Tassiulas, L.: Balanced RED: an algorithm to achieve fairness in the Internet. Proceedings of Infocom (1999)
31. Aweya, J., Ouellette, M., Montuno, D.Y.: A control theoretic approach to active queue management. Computer Networks, Vol. 36 (2001) 203-235
32. Floyd, S., Gummadi, R., Shenker, S.: Adaptive RED, An algorithm for Increasing the Robustness of RED's Active Queue Management. Submitted for publication. Available at http://www.aciri.org/floyd/

A Simplified Guaranteed Service for the Internet

Evgueni Ossipov and Gunnar Karlsson

Department of Microelectronics and Information Technology,
KTH, Royal Institute of Technology, Sweden,
{eosipov,gk} @imit.kth.se

Abstract. An expected growth of real-time traffic in the Internet will place stricter requirements on network performance. We are developing a new simplified service architecture that combines the strengths of the integrated and differentiated services architectures. In this paper we focus on the issues related to providing a guaranteed service in a high-speed network. We give a description of the service, which includes a lightweight signaling protocol and a non-work-conserving scheduling algorithm, and describe the system requirements and the performance evaluation. Our implementation of the protocol allows processing of 30 million signaling messages per second.

1   Introduction

Much effort today in the Internet research community is aimed at providing services for applications that were not under consideration when the Internet was originally designed. Nowadays the network has to support real-time communication services that allow clients to transport information with expectations on network performance in terms of loss rate, maximum end-to-end delay, and maximum delay jitter. A number of solutions to accomplish these needs have been proposed and implemented. The two most well-known approaches are the integrated services (*intserv*) [1] and differentiated services (*diffserv*) [2], [6] architectures. The intserv architecture classifies network traffic into three classes: *guaranteed service, controlled load service* and *best effort*. In diffserv, the traffic is assigned to specific behavior aggregates.

For traffic with guarantees, intserv provides reservation of bandwidth and buffers by using signaling between network nodes. An example of such signaling is RSVP [3]. Intserv keeps per-flow soft state in each network node. In comparison, diffserv avoids per-flow states in the routers, and instead ingress nodes perform traffic metering and admission control on the flows. The services offered to customers are statically described by service level agreements.

From the point of view of performance and scalability, intserv appeared to be a too cumbersome architecture for high-speed IP networks. Basically, the performance is limited by a router's ability to process and maintain the set of per-connection states. The diffserv architecture, however, is free of these limitations since a router serves all flows of a behavior aggregate together. The high scalability of diffserv is accomplished by separating the operations performed at the borders of the network

G. Carle and M. Zitterbart (Eds.): PfHSN 2002, LNCS 2334, pp. 147–163, 2002.

from those performed in the core. There is alas an inflexibility in this approach, namely a service level agreement (SLA) does not allow the end user to dynamically change the service requirements, for example maximum allowed bit rate.

The limited scalability of the intserv approach and lack of dynamic resource allocation in the diffserv architecture have motivated researchers to work on the combination of the two QoS approaches, rather than to consider them separately. That work is basically going on along two tracks. Some groups try to make the two approaches interoperable, for example by classifying traffic into the intserv traffic classes at the network ingress and then mapping aggregated flows of the same class to certain behavior aggregates of the diffserv model [4], [5], [7]. Other groups are working on simplifications of the signaling protocol [8-13]. The recently organized IETF working group on Next Steps in Signaling (NSIS) [30] works on a specification of the next generation signaling protocol that will overcome problems of RSVP and diffserv's static SLAs; however, there is no solution proposed yet.

The work presented in this paper develops the ideas of a QoS architecture proposed by Karlsson and Orava in [14]. It introduces a new simplified service architecture that combines the strengths of the intserv and diffserv models. Our contribution is as follows. We provide a description of the guaranteed service model, which includes a lightweight signaling protocol. We suggest a simple way of flow description based on a fixed rate. We further evaluate the non-work-conserving scheduling introduced in [15], and provide formal and experimental justification of its suitability for our service architecture.

The remainder of the paper is organized as follows. Section 2 describes the service architecture and its assumptions. In Section 3 the description of the simplified signaling is presented. Section 4 describes the scheduling algorithm with its system requirements, performance parameters and simulation results. We state open issues and summarize our work in Section 5.

2   Description of the Service Architecture

We are developing a service architecture consisting of three quality classes, as illustrated in Fig. 1. The *guaranteed service* (GS) class provides deterministic guarantees, namely constant throughput, absence of packet loss due to queue overflows in routers, and tightly bounded delay. Flows of the *controlled load service* (CLS) class have bounded packet loss and limited delay. The third class of service is the customary *best effort* (BE). The flows of this class have no guarantees, they obtain leftover capacity in the routers, and may be lost due to congestion. Both GS and CLS are allocated restricted shares of the link capacity so that BE traffic always has a possibility to get through. Best effort traffic is also using all unused portions of capacity reserved for CLS and GS traffic. Since GS and CLS are isolated from each other, we consider only the interaction between GS and BE in this paper. The work in our group that is related to CLS is described in [16], [17].

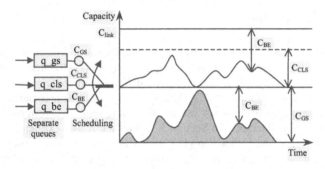

Fig. 1. The service architecture.

2.1   Assumptions Used in This Paper

For the purpose of our presentation we define the following actors in the network:

End systems (ES). These are client systems, which use certain services in the network, and the servers, which offer the services to the clients. We refer to data of a session between two end systems as a *GS flow*. Sufficient capacity for a particular GS flow is provided by explicit reservation in all nodes along a path. The ES's use a simple signaling protocol to make reservations. It is described below.

Access gateways (or ingress nodes) (AG). These are routers situated on the border between the backbone and the access networks of an ISP. An AG is the first IP-level node of a GS flow. In our architecture AG's keep per-connection soft state, perform policing and billing.

Core nodes (or routers). These are all routers apart from AG's. They provide IP-level packet forwarding, and do not keep per-connection soft state. Packets belonging to the GS class are queued in a separate GS queue at each node in the network, as shown in Fig. 1. The total traffic of all GS *flows* that share a particular link is referred to as an *aggregated GS flow*.

We base this work on the following assumptions.

- All nodes in the network are output-buffered devices that implement the scheduling proposed later in the paper (we are working on the issues related to the scheduling inside the switching fabric).
- Each GS connection is described solely by a peak rate as traffic specification. The signaling protocol assumes the reservation state to be purely additive: The reserved rate of an outgoing link of a router equals to the sum of the incoming rates for the link. This is enforced by the scheduling.
- We also assume that re-routing does not occur in the network. Although we are aware of the problem that routes in the real Internet may change, we leave this problem for our future work.
- We consider only unicast communications.
- Security, multicast, policing and billing are out of scope for this paper and will be considered in our future work.

3   Signaling Protocol

The signaling protocol serves the purpose of establishing a connection between two end systems by reserving capacity in the routers along a path. Related ideas have been proposed in [3], [8-13], [28]. Our protocol differs from these approaches in the following ways: it uses simpler traffic descriptor (we describe a connection solely by a peak rate); the protocol does not use per-flow soft state in core routers and relies instead on garbage collection (see Section 3.3). Further, it uses explicit reservations in comparison with [9]. In contrast to [11], which is similar to our approach in that it uses hard state, our protocol requires fewer messages to establish and maintain reservations; thus, our protocol introduces less overhead. The comprehensive work in [13] as well as efforts of ATM Forum [28] deal with signaling for QoS routing, while our protocol uses existing routing algorithms. Another approach to resource provisioning deals with centralized schemes (bandwidth brokers), like in [6], which is not considered in this paper.

We have the following design goals for our signaling protocol: It should not require per-connection soft-states in the routers; the number of signaling messages should be low, and the time to process a signaling message should be small. The protocol should be robust and tolerate loss of certain types of messages.

The idea behind the signaling is simple and uses the additivity of the reservation state. We keep track of available capacity for each network link in a variable C_{GS}, and have two public state variables for *requested* and *confirmed* capacity, R_r and R_c. All users are able to modify these variables by means of the signaling protocol. To make a reservation the router simply adds the requested rate in the signaling message to the requested capacity variable. This permits the requestor to send data with this rate. The router does the reverse operation to release capacity when the end system terminates the connection by an explicit tear down message. Let us define the requested rate r as a multiple of Δ, where Δ bits per second is a minimum value (quantum) of reservation. The following messages are needed for the signaling:

- Reservation message $mR(r)$ for the amount of capacity r. Messages of this kind may be discarded by the routers.
- Confirmation message $mC(r)$ from the initiating party for the reserved capacity. Messages of this type may not be lost.
- Tear down message $mT(r)$, which indicates that the initiating party finishes the connection and releases the reserved capacity. Messages of this type may not be lost.

The continued work on the protocol specification includes finding the best possible way of encoding the signaling messages in the fields of the IP header. We believe that the length of a signaling message could be as small as an IP header, 20 bytes, considering that we only need to represent three distinct messages and one rate value.

The service specifies that all signaling messages are carried in the GS part of the link capacity, and they are therefore not subjected to congestion-induced loss.

3.1   End System Behavior

In order to establish a guaranteed service connection, the initiator sends $mR(r)$ towards the destination. Each reservation message has a unique identifier. After issuing this message a waiting timer is started with a value T. This is the allowed waiting time for an acknowledgment at the sender and it is a constant specified for the service. By defining T as maximum RTT we ensure correct handling of delayed acknowledgments: If no acknowledgment arrives during T seconds, the initiator sends a new reservation request.

An acknowledgment message is issued by the receiver when it receives a reservation request, and it is encoded as a zero-rate reservation (i.e. $mR(0)$) with the same identifier as in the reservation message. For a sender an acknowledgment indicates a successful reservation along the path. The source then sends a confirmation message towards the destination, and may immediately start to send data at a rate below r b/s. When the initiating party wishes to end the session, it sends a teardown message, freeing the reserved capacity on the path. The diagram in Fig. 2 represents the behaviors of the sender and the receiver.

The establishment of a connection is repeated until an acknowledgment arrives or a sender decides to stop. If an acknowledgement was delayed in the network and the sender times out before receiving this message, it will issue a new reservation message. The acknowledgment for the old reservation will be ignored. The core routers will handle this case as a lost reservation, and will clean up the reserved capacity during the garbage collection (see Section 3.3).

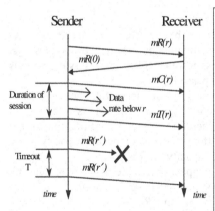

Fig. 2. Setup and tear down of reservations.

```
void server() {
    //types of messages: 0- mR, 1- mC, 2-mT
    r=0; //requested capacity
    Rr=0;//reserved capacity
    Rc=0; //confirmed capacity
    while (1) {
        int type=receive();
        switch type {
            0: if(r>Cgs-Rr) //Admission control
                drop(message);
            else Rr +=r
            1: Rc +=r;
            2: Rr-=r; Rc-=r;
        } //switch type
        if (Rr == Cgs) {
            wait(W);
            /* function wait process only messages
            of types 1 and 2 for the duration of W
            seconds */
            if (Rr!=Rc) garb_collect();
        } //if (Rr == Cgs)
    } //while (1)
} //void server()
```

Fig. 3. Algorithm for processing signaling messages in a router

3.2   Router Behavior

First when a router receives $mR(r)$ it performs an admission control based on the requested rate. The router will drop a reservation if it exceeds the remaining capacity $(r > C_{GS} - R_r)$. Otherwise it increases R_r by the bit rate requested in the reservation message. When getting a confirmation message, the router increases the value of R_c

by the rate r. When the router gets a teardown message, it decreases both R_r and R_c by the rate in the message. The summary of a router's behavior is presented by the pseudo-code in Fig. 3.

As one could notice from the algorithm, if a router gets an acknowledgment message in the form of $mR(0)$, it will not change the state variables. However, acknowledgments are carried in the unallocated part of the GS capacity, and might therefore be lost due to buffer overflow if that part is nil. Fig. 4 shows the capacity allocation for the signaling messages and the dynamic behavior of R_r and R_c. The message $mR(r)$ is carried in the available capacity from which it reserves a share; $mC(r)$ and $mT(r)$ use the established reservation. Thus, the messages are not subjected to congestion.

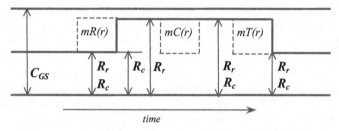

Fig. 4. Dynamic of R_r and R_c.

3.3   Garbage Collection

The reservation message is discarded when a router blocks a reservation request. This means that there will be pending reservations upstream along the path of the message. Thus, we need some garbage collection process to re-establish the reservation levels in the routers that the message has successfully passed.

Since a message of the type $mC()$ will not be lost, all successfully acknowledged reservations are expected to be confirmed. The garbage collection starts when all GS capacity is reserved $Rr=C_{GS}$. The router will therefore expect a confirmation for the reserved capacity within a waiting time $W>T$. If some of the requested capacity is not confirmed after W, the output port will free the capacity $Rr-Rc$.

Example. Consider the following case. The capacity variable of the outgoing link is $C_{GS}=5\ Mb/s$. There was no reservation request before T0, so $Rr=Rc=0$.

Table 1 shows the dynamic behavior of C_{GS}, R_r and R_c. Steps T1 to T3 show a successful cycle of a reservation for the amount of 1 Mb/s. The router gets a confirmation message at T2 that makes R_r and R_c equal. This indicates success of the reservation. A tear down message arrives at time T3. At T4 we have a 4 Mb/s requested reservation, and so far have only 2 Mb/s been confirmed. At time T5 the router accepts a 1 Mb/s reservation. Since the amount of requested capacity becomes equal to the available capacity C_{GS}, the garbage collection is started. During the time from T5 to T6, which equals W seconds, the router accepts only confirmation and tear down messages; all reservation messages will be discarded. There are several possibilities of what messages may arrive to the router during the garbage collection. One of the cases will be if the router gets confirmations for all unconfirmed capacity.

This indicates that all the capacity on the link is in use and the garbage collection is ended without action. Another case is when some or no confirmations will arrive. In this case the garbage collection procedure at time T6 will realize that the amount of capacity R_r-R_c= *5-2 =3 Mb/s* is not confirmed and will free it. Then the router will start accepting new reservations again.

Table 1. Bookkeeping of parameters for reservations.

Time Point	Event	C_{GS}	R_r	R_c
T0	Initial settings	5	0	0
T1	mR(1) arrives	5	1	0
T2	mC(1) arrives	5	1	1
T3	mT(1) arrives	5	0	0
....				
T4	Settings after some time	5	4	2
T5	mR(1) arrives	5	5	2
T5'	Garbage Collection starts [Accept only mC() and mT()]			
T6	Settings after Garbage Collection [Accept all messages]	5	2	2

The interesting case is when a teardown message will arrive during the garbage collection. Suppose during T5 to T6 the router has processed two 1Mb/s tear down messages.

Table 2. Arrival of tear down during garbage collection.

Time Point	Event	C_{GS}	R_r	R_c
T4	Settings after some time	5	4	2
T5	mR(1) arrives	5	5	2
T5'	Garbage Collection started [Accept only mC() and mT()]			
T1	mT(1) arrives	5	4	1
T2	mT(1) arrives	5	3	0
T2'				
T6	Settings after Garbage collection [Accept all messages]	5	0	0

In Table 2 the changes are reflected. We will denote the time points within this interval with small letters. At t2' we have 2 Mb/s of capacity which can be used by new reservations. One possibility is to stop the garbage collection and accept new reservations. However it is not efficient: Potentially it may result in the situation where the garbage collection restarts every time a new reservation arrives. Thus we have chosen to let the garbage collection end before accepting new reservations. Note that this choice affects only the duration of the blocking periods, but since it releases more capacity could result in fewer activations of the garbage collection.

3.4   System Requirements and Performance Parameters

The routers need to provide enough buffer space for the signaling messages that cannot be lost. We have assumed that there is a minimum reservation quantum. The worst case is when all capacity available for the guaranteed service will be reserved by the smallest quantum. In this case the total number of supported connections will be equal to C_{GS}/Δ. Since tear down and confirmation messages can be issued only for successful reservations, the total number of loss-intolerant messages will be twice the number of connections. If we assume that the call processing is equally fast in all routers then it is easy to show that a router with n ports requires $Bmax$ packets of buffering per output port:

$$B_{\max} = \frac{2C_{GS}(n-2)}{\Delta(n-1)} \,. \tag{1}$$

(The buffer size can be reduced in half if the end systems do not send confirmation and tear down messages back to back. This is accomplished by enforcing a minimum duration of a GS connection.)

We have implemented the signaling protocol in the network simulator ns-2 [29]. Our goal was to observe the performance of our protocol in a real router; therefore, we measured the actual time the processor spends on processing every signaling message and not the local ns-2 time, since it is independent of the computer power. In a series of simulations on a processor Intel Pentium III 500 MHz we obtain the following results. The time of processing one reservation message is 35 microseconds, a tear down message takes 32 microseconds, and a confirmation message 17 microseconds. Although the operations of reservation and tear down are similar, recall that the router performs admission control for reservations. This explains the slightly different processing times for these types of messages. The average rate of message processing is therefore 30×10^6 messages per second. Assuming for example a link capacity equal to 100 Mb/s and a reservation quantum of 10 kb/s, then the maximum number of supported connections will be 10000 and therefore a buffer for 20000 messages is needed to hold loss-intolerant signaling messages. This corresponds to 400 kB of memory if each message is 20 bytes.

Summarizing the description of our signaling protocol, we have developed a new simple signaling protocol for the guaranteed service. This protocol has the following properties.

- It is sender oriented.
- It has three types of messages $mR()$, $mC()$ and $mT()$ and uses four messages to setup, acknowledge, confirm and tear down a connection (with soft state, the number of messages depends on the connection's duration).

- The protocol does not use per-flow state in core routers and the simple operations allow fast message processing.

Further work on this signaling protocol will focus on the following problems:

- Encoding of signaling messages in standard IP header.
- Change of routes in the core network.
- Failure of a connection without explicit tear down.
- Loss of teardown or confirmation messages due to link failure or bit errors.

4.   Packet Scheduling in Routers

Many researchers work in the field of scheduling algorithms. A good overview of the existing approaches is given in [26]. Work-conserving algorithms like those in [20], [21] have been extensively studied in [22-25]. Non-work-conserving schemes are described in [18], [19], [27]. In our architecture we use the non-work-conserving scheme described in [15]. Our contribution described in this section is a further evaluation of the chosen scheme, with experimental and formal justification of its suitability in our service architecture.

In order to satisfy the properties of our service model, a router needs a special scheduling algorithm to enforce additivity of the reservation state and to ensure absence of packet loss for the GS traffic in the network. It should work with variable length packets up to some maximum size L. The algorithm should not prevent best effort traffic from being sent for long periods. The buffer space for the GS traffic in the routers should be enough to avoid packet loss, taking into account the worst arrival pattern.

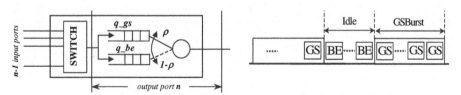

Fig. 5. An output port of a router.   Fig. 6. Scheduling.

4.1   The Algorithm

Consider a router in the network as shown in Fig. 5. The router has n ports. Traffic for output n enters the router through $n-1$ input ports. After the route lookup, packets come to the switching fabric and are switched to the appropriate output ports. At an output port, all incoming packets are sorted according to the service class and placed into the corresponding queue: one for guaranteed service (q_gs) and one for best effort service (q_be).

The scheduler will serve one or more packets from the GS queue (a *GSBurst*), then it will schedule an idle period long enough to preserve the outgoing reserved rate of GS traffic. During the idle it will serve packets from the best effort queue. Note that proposed scheduling scheme is non-work-conserving only with respect to GS packets. Best effort traffic gets all the capacity that is not used by GS. The schematic in Fig. 6 shows this scheduling. Thus we define the reservation ratio of an outgoing link as

$$\rho = \frac{GSBurst}{Idle + GSBurst}. \tag{2}$$

For a GS burst the duration of the idle period is

$$Idle = \frac{1-\rho}{\rho} GSBurst. \tag{3}$$

Note that with increasing ρ up to 1 the idle period tends to 0. Under such circumstances BE traffic will be blocked. Thus, we compute the smallest length of a *GSBurst* so that ρ is maintained and the following idle period is sufficiently long to transmit one BE packet of the maximum size L. Thus, the burst size is at most $\frac{\rho}{1-\rho} L$ bytes long. The pseudo-code of the scheduling algorithm is presented in [15].

The system in Fig. 5 has a drawback; namely, having our scheduling only at an output port, we cannot control the number of packets that might bunch together due to the multiplexing of flows from the input ports. The outgoing aggregate is smooth but the output scheduling does not control the behavior of each subaggregate.

Although the aggregate flow from a router does not contain bursts, we do not have information about how individual flows will be routed in the next router

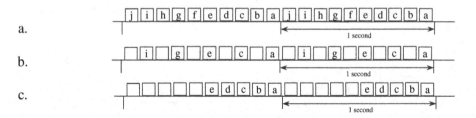

a.

b.

c.

Fig. 7. Burst of a subaggregate.

In Fig. 7a we have an aggregate flow of 10 packets per second (pps), observed at the output of a router during 2 seconds. The aggregate consists of 10 individual connections, each with rate 1 pps. All the connections are smooth. Now, consider this aggregate arriving to the demultiplexor of the next router. Assume that 5 pps of this aggregate should be forwarded to output port 1. The best case for the corresponding queue will be when the subaggregate of 5 pps will be smooth. This will happen if connections a,c,e,g,i are destined to port 1 (Fig. 7b). But it could also happen that the 5pps subaggregate will consists of flows a,b,c,d,e (Fig. 7c). The worst-case subaggregate is formed when the aggregate consists of connections with the minimum reservation rate.

We can eliminate this problem by adding similar schedulers at each input port. Their purpose is to smooth out the distortion of the subaggregates introduced by the upstream router. This will allow us to dimension the buffers needed in the routers for zero packet loss.

4.2   Buffer Size Evaluation

We add (n-1) rate controllers at each input port as shown in Fig. 8a. Each controller is responsible for smoothing out the subaggregate directed to a particular output port. Denote the controller at the input port i for the output port j as the (i,j)-controller. The rate controller here is the same scheduler as described in the previous section. At each output port there will be a scheduler for the purpose of shaping the multiplexed flows from all the input ports, and for providing a fair service to the best effort traffic. Hence from the point of view of an output port the architecture looks as in Fig. 8b.

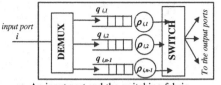
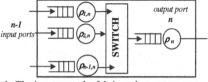

a. An input port and the switching fabric.   b. The input port, the fabric and an output port.

Fig. 8. Router architecture (GS-aware devices).

In this architecture we need a small buffer at the output, only $(n-2)L$ bytes long, in order to keep packets arriving simultaneously from the input ports. By having the rate controllers at the input we will never have bursty arrival to the queue of the output port. Now we have obtained the following:

- There is no clumping of packets from a particular input port.
- An aggregated output flow is smooth.
- While certain subaggregates of the output flow can be bursty, the burst is finite and easy to compute.

Having a rate controller at the input ensures the first property. Although a subaggregate can be distorted at the output due to multiplexing with flows from other input ports, it will be reshaped at the input of the downstream router. The presence of a non-work-conserving scheduler at the output ensures the second property. The observations about the third property were made in section 4.1.

We will compute the needed buffer for one (i,j)-controller at the input. Denote the reservation level for the flows originated from the port i on the outgoing link j as $\rho_{i,j}$. The rate of the (i,j)-controller is $\rho_{i,j}C_{GS}$, the minimum reservation rate is Δ, and the capacity available for GS is C_{GS}. Recall that in the worst case all the connections in the subaggregate will have minimum reservation rates. The number of connections will then be $N=\rho_{i,j}C_{GS}/\Delta$. The worst-case burst of the subaggregate arriving to $q_{i,j}$ is N packets with the rate of the aggregate (as in Fig. 7c). For the maximum size packet L, the needed buffer is:

$$B(\rho_{i,j}) = L \frac{C_{GS}\rho_{i,j}}{\Delta}(1 - \rho_{i,j}). \qquad (4)$$

As an example we have computed the size of one controller queue in the router with the following parameters. In the attached links 100 Mb/s is available for the guaranteed service (C_{GS}=100Mb/s), the minimum reservation rate Δ is 10 kb/s, the maximum packet size L is 576 bytes. In Fig. 9 the size is shown of an input buffer depending on the level of reservation $\rho_{i,j}$.

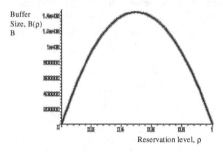

Fig. 9. Buffer size as a function of reservation level.

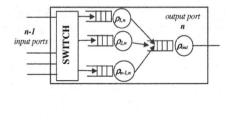

Fig. 10. Router architecture (implementation).

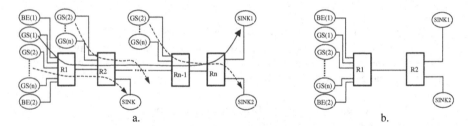

Fig. 11. Network topology for the simulations.

In the implementation we have moved all the rate controllers from the input ports to the corresponding output port. Fig. 10 shows this structure.

The scheme shown in Fig. 8 is functionally equivalent to the one in Fig. 10. In this work we are not concerned with the scheduling for the switching fabric. The only purpose of the controllers at the input is to prevent bursty arrivals to the buffer at the output port. From Fig. 9 one can see that the maximum buffer size is needed when the rate controller is setup to 0.5 of the outgoing link. In our example we need at most 1.4 MB of buffer in this controller. Assume the router has 50 input ports. The maximum size buffer for output port n will be when the output rate will be equally distributed between all input ports. In this case the total memory needed for buffers at the output port is $(n-1)B(\frac{\rho_n}{n-1}) + (n-1)L - L$, where n is the number of ports in the router and $B(\quad)$ is given in Eq. 4. In our case it will be 5.7MB.

4.3   Simulation and Worst Case Performance Analysis

In this section we present simulations to evaluate the delay and fairness properties of our scheduling. Our scheme provides a fair service for best effort traffic even when the reservation level is higher than 50 percent. We compare the performance of our scheduler to a strict priority scheduler. Both schemes were implemented in the network simulator NS-2 (we are working on an implementation of the proposed architecture in a Linux-based router, and will report the results in our future publications).

We used the topology in Fig. 11a to study the delay properties of our scheme. The goal was to measure the end-to-end delay of one GS flow. The following setup was used: $R(1)$ to $R(N)$ are the routers, $BE(1)$ and $BE(2)$ are two best effort sources, $GS(1)$ is the monitored GS source, $GS(2)$ up to $GS(N)$ are background GS sources. All links in the network are 100Mb/s. A flow from $GS(1)$ traverses the path towards SINK1. The best effort flows from $BE(1)$ and $BE(2)$ follow the path towards SINK2. The background GS traffic follows the path towards a SINK connected to the next downstream router. We ran simulations with both VBR and CBR traffic. In the case of CBR traffic, the sources have the following characteristics: $GS(1)$ is set to 4Mb/s; the rest of the capacity available for the guaranteed flows is equally distributed between sources $GS(2)$ up to $GS(N)$; the best effort flow from $BE(1)$ is 10 Mb/s; the rest of the capacity available for the best effort flows is assigned to $BE(2)$. In the case of VBR traffic, all GS sources are exponential on-off sources. Their peak rates are the same as in the CBR case. The mean duration of on/off periods is 500 milliseconds.

Initially experiments were done with VBR traffic. The number of hops between $GS(1)$ and SINK1 was 3, 5, 10, and 15. We used 10 background GS sources in this experiment, feeding each router. The monitored flow of 4Mb/s from $GS(1)$ was subjected to the same kind of background traffic in every router along the path. We conducted two experiments, firstly with a reservation of 80 percent on all links, and second with 20 percent. The goal was to measure the end-to-end delay of the monitored flow. Then we repeated the simulations for the CBR traffic.

Fig. 12 shows the delay jitter (the difference between the maximum and minimum delays) of the monitored GS flow for the VBR case. Simulations show for both values of the reservation level that our scheme introduces much smaller jitter than strict priority scheduling. This is due to the difference between the two schemes. Using strict priority scheduling a burst of packets can be created due to the effect of flow multiplexing. This phenomenon may increase with number of hops. While in our scheme the multiplexed flows from the input ports in a router are reshaped and smooth. By looking at the maximum and minimum delays for our scheme we have seen that they grow almost linearly with the number of hops. Fig. 13 shows this. This explains the small variations in the delay jitter. In the case of $\rho=0.8$ the maximum jitter is 0.414 milliseconds after 15 hops, in the case where $\rho=0.2$ this value is 2.3 milliseconds. Reservation values of less than 0.5 (in our case 0.2), lead to packets being delayed for one or more idle periods. This accounts for on higher jitter value. We have studied the maximum delay under the worst case arrival of the GS traffic in CBR experiments. The maximum delay of the 4Mb/s after 15 routers with 10 input ports each is 11 milliseconds (excluding the propagation time on the links).

We decided to continue our comparison of the two schemes using a different topology. In this case we did not want to include the hop count. We studied the impact of the two schemes on a monitored best effort flow. For this purpose we used the topology in Fig. 11b. The description of the sources, routers, links as well as traffic specification is the same as in the previous setup. The monitored BE flow from BE(1) follows the path towards SINK1, and other flows towards SINK2. The number of GS sources was varied from 10 to 100. The results of the experiments with ρ=0.8 and ρ=0.2 are shown in Fig. 14. The following observation was drawn from the simulations: For both values of the reservation level our scheme performs better with regard to best effort flows. This is due to the difference between the two scheduling schemes: Under strict priority scheduling GS traffic gets full priority over best effort traffic, therefore the BE packets can only be transmitted when there are no packets in the GS queue, while in our scheme, the best effort traffic gets certain guarantees to be transmitted even in case of simultaneous arrivals of GS packets.

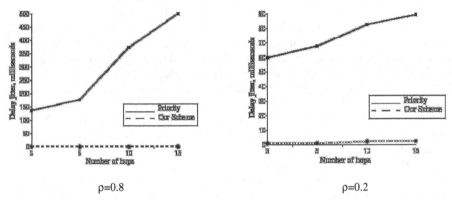

ρ=0.8   ρ=0.2

Fig. 12. End-to-end jitter of the monitored GS flow for the VBR case.

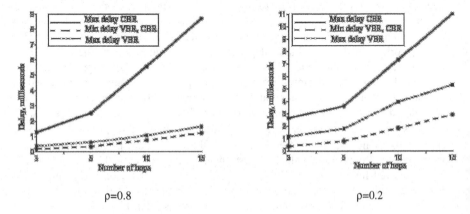

ρ=0.8   ρ=0.2

Fig. 13. Min and max delays for GS flow for VBR and CBR.

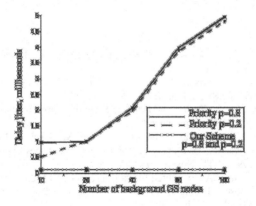

Fig. 14. Delay jitter of the monitored best effort flow.

5   Open Issues and Summary

The ongoing work on the architecture continues in the following ways. Current specification of the signaling protocol does not support loss of other kinds of messages than reservation messages. Since a teardown message might be lost, due to link failure for example, all the nodes after the point of failure will never receive the message and therefore the amount of capacity, which was supposed to be deallocated, will remain in use. This situation can potentially lead to a situation in which all the capacity in a router is unavailable for new reservations. To eliminate this we are developing a more advanced garbage collection, to increase the robustness of the protocol. We also develop a mechanism in the signaling protocol which will handle re-routing. Obviously forwarding rerouted GS traffic without reservation is unacceptable.

The design in this paper was concerned only with unicast communication. Another direction of our work is to consider multicast GS flows. We are working on an implementation of the described architecture in a Linux-based platform. We intend to consider the details of the administration policy, security and pricing issues of our service.

We have presented a simplified guaranteed service for the Internet. Our contribution is as follows. We suggest a simpler way of flow description based on a fixed rate. This along with a new signaling protocol and non-work-conserving scheduling gives a basis for capacity reservation and admission control. The simulation results show good performance of the service. The simplicity of the overall architecture certainly adds to the possibility of implementing it in high-speed routers.

References

1. R. Braden, D. Clark, S. Shenker, "Integrated services in the Internet architecture: an overview", RFC1633, IETF, June 1994.
2. S. Blake, D.Black, M.Carlson, E.Davies, Z. Wang, W. Weiss, "An architecture for differentiated services", RFC 2475, IETF, December 1998
3. R. Braden, L. Zhang, S. Berson, S. Herzog, S. Jamin, "Resource reservation protocol (RSVP)", RFC 2205, IETF, September 1997
4. Y. Bernet, P. Ford, R. Yavatkar, F. Baker, L. Zhang, M. Speer, R. Braden, B. Davie, J. Wroclawski, E. Felstaine, "A framework for integrated services operation over diffserv networks", RFC 2998, IETF, November 2000
5. V. Jacobson, K. Poduri, "An expedited forwarding PHB", RFC 2598, IETF, June 1999
6. K.Nichols, V. Jacobson, L.Zhang, "A two-bit differentiated service architecture for the Internet", RFC 2638, IETF, July 1999.
7. J. Harju, P. Kivimaki, "Co-operation and comparison of diffServ and intserv: performance measurements", *in Proc. of 25th Annual IEEE Conference on Local Computer Networks*, 2000, pp. 177 – 186
8. P. Pan, H. Schulzrinne, "YESSIR: a simple reservation mechanism for the Internet", *Computer Communication Review*, vol. 29, No. 2, April 1999
9. W. Almesberger, T. Ferrari, J.-Y. Le Boudec, "SRP: a scalable resource reservation protocol for the Internet", *Sixth International Workshop on Quality of Service (IWQoS 98)*, 1998, pp. 107 –116
10. G. Feher, K. Nemeth, M. Maliosz, I. Cselenyi, J. Bergkvist, D. Ahlard, T. Engborg, "Boomerang - a simple protocol for resource reservation in IP networks", *IEEE Workshop on QoS Support for Real-Time Internet Application*, June 1999.
11. A. Eriksson, C.Gehrman, "Robust and secure light-weight resource reservation for unicast IP traffic", *Sixth International Workshop on Quality of Service (IWQoS 98)*, 1998, pp. 168-170
12. T.W.K. Chung, H.C.B. Chang, V.C.M. Leung, "Flow initiation and reservation tree (FIRST)", *IEEE Conference on Communications, Computers, and Signal Processing*, pp. 361-364, 1999.
13. R.A. Guerin, A. Orda, "Networks with advance reservations: the routing prespective", *in Proc. of INFOCOM 2000*, pp. 118-127.
14. G. Karlsson, F. Orava, "The DIY approach to QoS", *Seventh International Workshop on Quality of Service (IWQoS 99)*, 1999, pp. 6 –8
15. M. Mowbray, G. Karlsson, T. Köhler, "Capacity reservation for multimedia traffics", *Distributed Systems Engineering*, vol. 5, 1998, pp. 12-18
16. V.Elek, G.Karlsson, R. Rönngren, "Admission control based on end-to-end measurements", *in Proc. of INFOCOM 2000*, pp. 623-630.
17. I. Mas Ivars, G. Karlsson, "PBAC: probe based admission control", *In Proc.of Second International Workshop on Quality of Future Internet Services (QoFIS 2001)*, pp.97-109, 2001.
18. S.J. Golestani, "A framing strategy for congestion management", *IEEE Journal on Selected Areas in Communications*, vol. 97, September 1991, pp. 1064 -1077.
19. R. Brown, "Calendar queues: a fast O(1) priority queue implementation for the simulation event set problem", *Communications of the ACM*, 31 (10), pp. 1220-1227, Oct. 1988.
20. J.C.R Bennett, H. Zhang, "WF^{2}Q: Worst-case fair weighted queueing", *In Proc. of INFOCOM 1996*, pp. 120-128.
21. S. Golestani, "A self-clocked fair queueing scheme for broadband applications", *In Proc. of INFOCOM 1994*, pp. 636-646.
22. R.L. Cruz, "A calculus for network delay. II. Network analysis", *IEEE Transactions on Information Theory*, Vol. 37, Issue 1, Jan. 1991, pp. 132-141.

23. R.L. Cruz.,"A calculus for network delay. I. Network elements in isolation", *IEEE Transactions on Information Theory*, Vol. 37, Issue 1, Jan. 1991, pp. 114-131.

24. L. Georgiadis, R.A. Guerin, A. Parekh, "Optimal multiplexing on a single link: delay and buffer requirements", *IEEE Transactions on Information Theory*, Vol. 43, No. 5, pp. 1518-1535, Sept 1997.

25. J.Y. Le Boudec, P.Thiran, *Network Calculus*, Springer Verlag LNCS 2050, June 2001.

26. H. Zhang, "Service disciplines for guaranteed performance service in packet-switching networks", *Proceedings of IEEE*, Vol. 83(10), pp. 1374-1396, Oct.1995.

27. H. Zhang, "Providing end-to-end performance guarantees using non-work-conserving disciplines", *Computer Communications: Special Issue on System Support for Multimedia Computing*, 18 (10), Oct. 1995.

28. ATM Forum, http://www.atmforum.com

29. Network Simulator ns-2, http://www.isi.edu/nsnam/ns/

30. Next Steps in Signaling (NSIS), IETF working group, http://www.ietf.org/html.charters/nsis-charter.html

Improvements to Core Stateless Fair Queueing*

Cristel Pelsser[1] and Stefaan De Cnodder[2]

[1] University of Namur, Belgium,
cpe@infonet.fundp.ac.be
[2] Alcatel, Belgium,
stefaan.de_cnodder@alcatel.be

Abstract. Core Stateless Fair Queueing (CSFQ) is a scalable mechanism to provide per-flow fairness in high-speed networks in that it does not need to maintain per-flow state in the core routers. This is possible because the state for each flow is encoded as special labels inside each packet. In this paper, we propose and evaluate by simulations two improvements to CSFQ. First, we show that CSFQ does not provide a fair service when some links are not congested. Our first improvement solves this issue. Second, we propose an algorithm to allow CSFQ to provide a service with a minimum guaranteed bandwidth and evaluate its performance with TCP traffic.

1 Introduction

The Internet was initially designed to provide a best-effort service. During the last ten years, two architectures have been proposed to allow the Internet to support other types of services. The Integrated Services architecture (IntServ) [2] aims at providing end-to-end guarantees to individual flows. The Differentiated Services architecture (DiffServ) [1] aims at providing several grades of service to different aggregated flows. Although these two architectures differ in most aspects, the routers supporting these advanced services will have to rely on classifiers, buffer acceptance algorithms, markers, traffic conditioners and schedulers to provide the required service differentiation or guarantees. The performance and the complexity of these mechanisms is a key issue to be considered when designing high speed routers.

From an application point of view, the IntServ architecture provides the best service since the guarantees are associated with each individual flow. However, this often forces the routers to perform a complex classification and to maintain some state for each individual flow. This severely limits the scalability of the IntServ architecture and its capability of being used in high speed networks. The DiffServ architecture on the other hand started from the assumption that it must be implementable with today's equipment. For this, DiffServ traded the per-flow guarantees in favor of per traffic aggregate guarantees. This improves the scalability of DiffServ. In DiffServ, a network is composed of two types

---

* This work was partially supported by the European Commission within the IST ATRIUM project.

G. Carle and M. Zitterbart (Eds.): PfHSN 2002, LNCS 2334, pp. 164–179, 2002.

of routers : edge routers and core routers. The edge routers typically serve a small number of customers with a few links. Their role is to classify the packets received from the customers and mark each packet with the class of service that they should receive. By putting the complexity at the edge routers this way, DiffServ is much more scalable and performant than IntServ but it lowers the quality of the service from the application point of view.

Inside high-speed networks, the performance of the routers is a key issue. Congestion should not occur at the routers. In other words, the routers should not be bottlenecks. They should be able to use all the bandwidth available on the links to which they are connected. As a consequence, the mechanisms that aim to provide per-flow guarantees should be able to work at a high pace, to be usable in high-speed networks. Therefore, Dynamic Packet State (DPS) [9, 11] is an interesting solution to provide per-flow service differentiation in a performant and scalable manner. DPS removes the requirement to maintain per-flow state by encoding the state for each flow as labels inside the IP packets. By this mean, it also removes the need for packet classification at each node. With DPS, as with DiffServ, there are edge and core routers. The edge routers classify the packets received from the customers and mark each packet with the per-flow state required by downstream routers to provide the required fairness or guarantees. Edge and core routers then process the packets according to these markings[1]. Since all the per-flow state information is included inside each IP packet, the core routers do not need to maintain per-flow state and to perform classification ; their complexity is reduced. To respect transparency, when a packet reaches the edge of the network, the packet state introduced by the ingress edge router is removed. Several mechanisms relying on the DPS principle or on the Differentiated Services Code Point (DSCP) have been proposed and analyzed in the literature [3, 4, 6, 7, 9, 10, 11, 12] for the provision of a large variety of services in high-speed networks. In this paper, we focus exclusively on Core Stateless Fair Queueing (CSFQ) due to space limitations. A more detailed overview of these mechanisms may be found in [8].

This paper is organized as follows. First, in Sect. 2, we provide a detailed description of CSFQ as it was proposed in [10]. Then, in Sect. 3, we show that CSFQ does not provide fairness when some links are non congested. We propose and evaluate a solution to this problem that occurs on many links inside high-speed networks. In Sect. 4, we show how to extend the CSFQ algorithm to provide a minimum guaranteed bandwidth to each flow and evaluate the performance of this extension with TCP traffic.

2   CSFQ Algorithm

By fairness we mean, in this paper, that all flows sharing a link, and having packets dropped at the router upstream of the link, should get the same amount of bandwidth. Such flows are said to be bottlenecked on the link. All other flows on the link get a smaller amount of bandwidth. The amount of bandwidth that

---

[1] For information concerning the storage of the label in the packets refer to [10, 11].

is allocated, in a fair situation, to a flow bottlenecked on a link is called the fair share of the link. In addition to the fair provision of bandwidth, this paper deals with the support of minimum bandwidth guarantees, allowing certain flows to use a portion of bandwidth in all case, in addition to their fair allocation.

CSFQ [10] aims at distributing bandwidth fairly between the flows sharing the network. It also allows to associate a weight to the different flows and tries to distribute the bandwidth in proportion to those weights. CSFQ is based on DPS. Since it does not need to perform classification, it is suitable for high-speed networks. The complexity is put into the edge routers, that are connected only to a few links, leaving the core routers with a lower complexity.

On packet p arrival, the edge routers perform classification. They determine the flow to which the packet belongs, $flow_i$. The state of this flow is then used to compute the arrival rate of the flow, AR_i. The arrival rate of $flow_i$ is determined for any i by

$$AR_{i,\text{new}} = (1 - e^{-\frac{T}{K}})\frac{L}{T} + e^{-\frac{T}{K}} AR_{i,\text{old}} , \tag{1}$$

where $AR_{i,\text{old}}$ is the previous estimation of the arrival rate of $flow_i$ stored in the flow state, K is a constant, L is the length of the current packet p and T is the time elapsed between the arrival of the previous packet of $flow_i$ and the current packet arrival. This is the formula used by the `estimate_rate` function in Fig. 1. The new arrival rate estimation of $flow_i$ ($AR_{i,\text{new}}$) is used to label the current packet p belonging to $flow_i$.

When a packet p arrives, the edge and core routers compute the drop probability of the packet using

$$P_{\text{drop}} = \max(0, 1 - \frac{FS}{\text{p.label}}) , \tag{2}$$

where the fair share, FS, has been estimated at the previous packet arrival and p.label is the label of the current packet. It can be deduced that, if p.label $> FS$

```
On receiving packet p
    if (edge router)
        i = classify(p);
        p.label = estimate_rate(AR_i,p);
    P_drop = max(0,1-FS/p.label);
    if (P_drop > unifrand(0,1))
        FS = estimate_FS(p,1);
        drop(p);
    else
            if (P_drop > 0)
            p.label = FS;  /* relabel p*/
        FS = estimate_FS(p,0);
        enqueue(p);
```

Fig. 1. CSFQ pseudo-code

then $P_{\text{drop}} > 0$. And, when p.label $\leq FS$, we have $\frac{FS}{\text{p.label}} \geq 1$. It follows that $P_{\text{drop}} = 0$.

Then, if the drop probability P_{drop} of the packet is greater than a random number, the packet is in excess of the fair share. A new estimation of the fair share is done and the packet is dropped. Otherwise, P_{drop} is lower than the random number and the packet belongs to the fair allocation. If the drop probability is above zero, the packet is relabeled with the fair share, which should be an approximation of the rate of $flow_i$ on the output link. Then, independently of the value of P_{drop}, the fair share is estimated. And, after the estimation, the packet is stored in the queue provided that it is not full. Figure 1 shows the pseudo-code of the CSFQ algorithm.

The estimation of the fair share is shown in Fig. 2. First of all, the arrival rate of the aggregate traffic, AR, at the router and the rate at which the packets are forwarded to the FIFO queue, FR, are estimated using the exponential averaging (1). The link is considered congested if the arrival rate is larger than the link rate (BW) and congested otherwise.

When the link becomes congested, the variable congested is set and, the beginning of the time window, start_time, is set to the current time. If the link remains congested for K_c seconds, the fair share is updated according to

```
estimate_FS(p,dropped)
    estimate_rate(AR,p);  /* estimate arrival rate */
    if (dropped == FALSE)
        estimate_rate(FR,p);
    if (AR≥BW)
        if (congested == FALSE)
            congested = TRUE;
            start_time = current_time;
        else
            if (current_time > start_time + Kc);
                FS = FS * BW/FR;
                start_time = current_time;
    else  /* AR < BW */
        if (congested == TRUE)
            congested = FALSE;
            start_time = current_time;
            temp_FS = 0;  /* Used to compute new FS */
        else
            if (current_time < start_time + Kc)
                temp_FS = max(temp_FS,p.label);
            else
                FS = temp_FS;
                start_time = current_time;
                temp_FS = 0;
    return FS;
```

Fig. 2. Fair share estimation pseudo-code

$$FS_{\text{new}} = FS_{\text{old}} \frac{BW}{FR} \ . \tag{3}$$

An explanation of (3) is given in [10]. Intuitively, it can be seen that when the forwarding rate estimation is above the link rate, the fair share estimation will decrease leading to less packets being transmitted when the number of flows stays constant or decreases. Otherwise, the fair share estimation increases.

The link is considered uncongested when the arrival rate is below the link rate. If the link was congested at the previous estimation but is now uncongested, the variable congested is set to false and temp_FS is set to zero. The variable temp_FS is used to compute the maximum of the labels carried by the packets arriving during an interval of length K_c and starting at start_time.

It can be noticed that, when the link isn't a bottleneck, there is no need to drop packets from any flow. In this case, the fair share is set to the arrival rate of the flow with the highest bandwidth. By setting the fair share at the maximum flow rate, no packets will be dropped because, according to (2), the dropping probability will be zero for each flow.

The fair share FS is estimated each time the link is congested for at least K_c seconds. It is also computed when the link is uncongested during the last K_c seconds. We can already see that if the link passes from uncongested to congested and vice-versa too often (i.e. in less than K_c seconds) the fair share will not be reestimated.

3   Improvement to CSFQ

3.1   Uncongested Network Problem

It has to be noticed that in CSFQ (Fig. 2), as exposed in [10], when there is no congestion, the fair share is set to the maximum of the packets' labels passing through the node during a fixed interval, called a window size, of length K_c. But, during such intervals, sometimes not all flows have packets arriving at a node. Because TCP is bursty, the rate of a TCP flow may be high even if no packets are sent during certain periods. Consequently, the fair share might be smaller than the label of certain packets. As a result, the drop probability of certain packets could be higher than zero. Some packets may be dropped even if the link on which they have to be transmitted is not congested.

Another problem with the estimation of the fair share as the maximum of the labels, in uncongested mode, is that the flows have difficulties to increase their sending rate. For example, if the maximum arrival rate among the flows at a link increases, and the new arrival rate of the flow is not yet incorporated in the fair share estimation, some packets of the flow are dropped. Again, we observe dropping of packets during an uncongested period. But, during uncongested periods, TCP always tries to increase its sending rate. If packets of the TCP flows are dropped, their sending rate will decrease because dropped packets are interpreted as a congestion indication by TCP. Therefore, this estimation of the fair share done by CSFQ for uncongested links can be seen as unfriendly toward

TCP flows. Additionally, forcing the flows to decrease their sending rate leads to an under-utilization of the resources.

We can also imagine the case where a network node does not have any packet to send on a particular output link. Then, the estimation of the arrival rate should be around 0 bps and the link is considered as uncongested. Because no packets arrive, the fair share is estimated to 0 bps. And, when packets will arrive for this link, they will all be dropped. If the link stays uncongested during a window size, the fair share estimation will increase to the maximum of the labels that passed during the last window size period but if the link becomes congested before this update, the fair share estimation will remain zero according to (3).

3.2 Proposed Solution

To avoid the dropping of packets when there is no congestion (Sect. 3.1), we propose to estimate the fair share on the entire period that the network is uncongested, instead of during K_c seconds. And, the fair share is still updated every K_c seconds. This means that the last estimation of the fair share is taken into account in the new estimation. Instead of computing the maximum starting from 0 at the beginning of the K_c windows, the maximum is computed starting from the previous fair share estimation. In other words, every K_c seconds an estimation is made over the last K_c seconds, and the final estimation is set to the maximum of the current estimation and the new maximum, calculated over the last period. Figure 3, shows the pseudo code of the fair share estimation in our Modified version of Core Stateless Fair Queueing (MCSFQ). This requires only changes in the non congested case of the fair share estimation.

The solution proposed is interesting because, when the link becomes uncongested, it can be considered that the right fair share has been found. The new fair share should not be lower than the fair share used when there was congestion because the arrival rate decreased, so, less packets have to be dropped. It is logical that the fair share should be the maximum of the fair share at the last congestion update and the packet's labels that passed. When the link stays uncongested, the temporary variable is not reset because all packets can be forwarded. When resetting it, like in the proposal made in [10], if the labels of the packets that pass during a period of a window size length are strictly lower than the maximum of the packets' labels passed during the previous period, the fair share decreases. But, when there is no congestion, either each flow has already reached its fair share or the flows are not using all network resources. There is no need to limit the use of these resources.

Now, during uncongested periods, the fair share is estimated by the maximum of the packets' labels, received since the end of the congestion and of the last fair share update before the uncongested period. It follows that the fair share estimation is higher with MCSFQ than with CSFQ, allowing to forward packets at a higher rate. Additionally, TCP will more easily be able to increase its sending rate. When TCP increases its sending rate, the first packets with a high label may be dropped but their labels are taken into account in the next fair share

```
estimate_FS(p,dropped)
    estimate_rate(AR,p);   /* estimate arrival rate */
    if (dropped == FALSE)
        estimate_rate(FR,p);
    if (AR≥BW)
        if (congested == FALSE)
            congested = TRUE;
            start_time = current_time;
        else
            if (current_time > start_time + Kc);
                FS = FS * BW/FR;
                start_time = current_time;
    else   /* AR < BW */
        if (congested == TRUE)
            congested = FALSE;
            start_time = current_time;
            temp_FS = FS;   /* Used to compute new FS */
        else
            if (current_time < start_time + Kc)
                temp_FS = max(temp_FS,p.label);
            else
                FS = temp_FS;
                start_time = current_time;
    return FS;
```

Fig. 3. MCSFQ fair share estimation : pseudo-code

updates. It follows that TCP is able to send at a higher rate than before once its window size has increased again, when the network stays uncongested.

It can be noticed that, when the link becomes congested for a window size period, the fair share will be updated according to (3). If the fair share is too high when the link becomes congested, the forwarding rate measured during the first window period will be higher than the link rate and therefore, the next estimation of the fair share will be smaller.

3.3   Simulations

In each of the scenarios used for the simulations[2], there is one flow per source. These flows may be an aggregate of TCP or UDP flows. The data packets are sent from source to destination. The data packets all go in the same direction, from the left to the right of the networks. And, the acknowledgments follow the reverse path. Two way traffic, i.e. traffic with a mix of acknowledgments and data packets, is not considered in this paper. Because the acknowledgments are much smaller and less frequent than the data packets, there is never congestion in the reverse path. In the simulations, CSFQ (MCSFQ) is not used on the path of the acknowledgments. Only the data path is influenced by CSFQ (MCSFQ).

---

[2] OPNET is the tool used to perform the simulations.

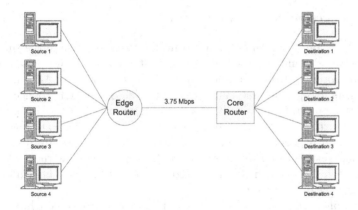

Fig. 4. Single bottleneck scenario

The bandwidths that are indicated on the Fig. 4 and Fig. 8 are the bandwidths for the data path. For the other direction, the link rates are set large enough such that there is never congestion. The simulations are done with a threshold[3] of half the queue size, as recommended in [10]. And, the fair share is initialized at 1 bps.[4]

3.4 Single Bottleneck Scenario

Although the single bottleneck scenario is simple, it already gives a good indication of the behavior of CSFQ (MCSFQ). It indicates whether or not CSFQ (MCSFQ) is able to distribute the bandwidth in a fair way in simple networks.

In the single bottleneck scenario (Fig. 4), there are four sources, an ingress edge router, a core router and four destinations. All packets generated by one source belong to the same flow, so there are four flows in total. All four flows cross the two routers of the network. They first pass through the edge router where they are multiplexed on a single link on which CSFQ (MCSFQ) allocates the bandwidth. Then, the flows cross the core router. Finally they are distributed on the links leading to their destinations. The edge router is the only possible bottleneck. The core router performs CSFQ (MCSFQ) also, but the traffic in excess to a fair bandwidth allocation should already have been dropped by the first router.

In this scenario, the links from the users to the routers have a rate of 100 Mbps with a fixed propagation delay of 2.5 ms. The link joining the two routers is of 3.75 Mbps with a fixed propagation delay of 10 ms in each direction.

---

[3] The threshold is a value used in case the network is uncongested. The network is supposed to stay that way until the queue occupancy gets above the threshold.

[4] For a discussion concerning the importance of the value chosen for the fair share at initialization refer to [8].

3.5   Uncongested Network Problem

In this subsection, the results, obtained from simulations where the single bottleneck network is uncongested or only congested at times, are presented. The results are obtained from statistics taken at the edge router. All sources send UDP packets except the first source that establishes one TCP connection. The UDP flows are not responsible of a possible congestion. Their rate of 500 Kbps is below one fourth of the link bandwidth. The total rate of the UDP flows will be 1.5 Mbps. That means that 2.25 Mbps are left to the TCP flow. In this situation, the fair share estimation should be around 2.25 Mbps to allow the TCP flow to make use of the bandwidth that is unused by the UDP flows. The window sizes, K_c, are first set to 0.6 seconds.

The graphs on the left of Fig. 5 and Fig. 6 show the throughput of each flow at the edge router with CSFQ, MCSFQ and also with tail drop as the only bandwidth allocation mechanism. The graphs on the right of these figures illustrate the dropping rate of the flows with the same bandwidth distribution mechanisms.

On Fig. 5 (left), it can be seen that the throughput of the UDP flows is around 500 Kbps independently of the bandwidth allocation mechanism. On the other hand, the throughput of the TCP flow varies depending on the bandwidth distribution mechanism used. It is at its peak with tail drop, it is slightly lower

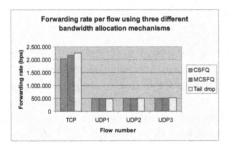
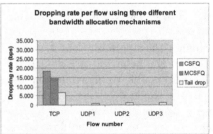

Fig. 5. MCSFQ versus CSFQ : $K_c = 0.6$s

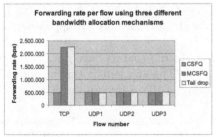
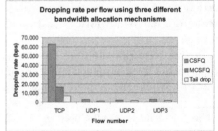

Fig. 6. MCSFQ versus CSFQ : $K_c = 0.1$s

with MCSFQ and it is the smallest with CSFQ. This means that the bandwidth resources are better used with tail drop than with MCSFQ. And, even less resources are used with CSFQ because the estimated fair share is higher with MCSFQ than with CSFQ in uncongested periods. When tail drop is used, UDP packets may be dropped as well as TCP packets (Fig. 5, right), which is not a fair behavior in this situation. With MCSFQ and CSFQ, the fair share estimation is at least equal to the rate of the UDP flows. It follows that no UDP packets are dropped (Fig. 5, right).

When the same simulations are performed with the window sizes set to 0.1 seconds [10] instead of 0.6 seconds, the TCP flow is only able to use the same portion of bandwidth as the one used by each UDP flow (Fig. 6, left), if CSFQ is used. Here, the window size is rather small and because TCP is bursty, its packets carry labels that are above the fair share estimation. Therefore, some TCP packets are dropped even if there is no congestion on the link. On Fig. 6 (right), it can be seen that even UDP packets are dropped that way. This sensitivity of CSFQ to the value of K_c is annoying in practice. Our Modified CSFQ does not suffer from this problem.

The TCP flow is not able to send above the throughput of the UDP flows with CSFQ when the window size is small. As exposed in Sect. 3.1, the TCP flow has difficulties in increasing its sending rate. With a higher window size, the fair share estimation is higher and there are more chances for the TCP flow to be able to increase its sending rate. The MCSFQ bandwidth allocation mechanism is not subject to such an important impact of the window size on the throughput of the flows in uncongested networks (Sect. 3.2).

On Fig. 6 (right), it can be seen that packets from the UDP flows are dropped with tail drop and CSFQ. We cannot say that this is a fair behavior because these flows do not create the congestion. On the contrary, no UDP packet is dropped by MCSFQ. The throughput of TCP is almost the same with MCSFQ and tail drop. Here we can say that MCSFQ is the fairer bandwidth allocation mechanism. We notice that MCSFQ is less sensitive to the window size than CSFQ. If the window size is small the bandwidth is not shared fairly with CSFQ as seen in Fig. 6. And, a high window size does also not provide fairness as shown by [8]. It is interesting to use a small value for K_c to have an estimation of the fair share that is more reactive to the changing traffic patterns. With a small window size, the fair share is estimated more often.

4   Support of Minimum Guarantees

CSFQ was conceived to distribute the bandwidth fairly among the flows sharing the network. It was thought to support flows that have different weights. The bandwidth that each flow receives is then a proportion of its weight. But, CSFQ was not meant, in [10], to be able to provide minimum guarantees to some flows. In this section, we show how to support minimum guaranteed flows in CSFQ and MCSFQ.

A small change in the labeling performed by the edge routers is enough to support minimum throughput guarantees. Some modifications may be done to the buffer acceptance module also [8]. But, these alterations are not mandatory to obtain satisfying results. In addition to the enhancement to CSFQ, an admission control mechanism[5] should be implemented for the support of minimum guaranteed flows to ensure that there are enough resources to provide the guarantees associated to the flows accepted in the network.

Our proposed modification to CSFQ is as follows. First of all, while labeling packets, edge routers have to be able to determine which packets are part of the guarantee associated to the packet's flow and which packets will be treated as best-effort. Bandwidth that is not part of any reservation has to be shared fairly between the flows sending packets in excess of their traffic contract. Therefore, CSFQ will only apply to excess packets. Edge routers will for that purpose mark all guaranteed packets with a label of zero. This indicates that the packet doesn't use any portion of the bandwidth that has to be allocated fairly. But, in excess packets will be marked with a label equal to the estimated excess rate of the flow, as suggested in [6]. A similar principle is also used in [12] for the provision of minimum throughput guarantees by Stateless Prioritized Fair Queueing (SPFQ).

When a packet arrives at a core router, it is checked, to determine the packet's dropping probability (2), if its label is higher than the fair share of the output link on which it has to be transmitted. When the packet is guaranteed, the fair share will never be smaller than the label. The drop probability of the packet is set equal to zero and the packet is not discarded, except if the queue is full and tail drop occurs. But, the drop probabilities of excess packets might be greater than zero.

The changes required to provide flows with minimum throughput guarantees are shown in Fig. 7. In the pseudo code given in Fig. 7, $guar\_rate_i$ is the amount of bandwidth that is guaranteed to $flow_i$ and ER_i is its excess rate. The marking is performed based on a probabilistic determination of packets in and out of the guarantees. A deterministic method may also be considered by using token buckets. The excess rate of the flow is estimated according to (1). Moreover, to obtain a quicker convergence of the fair share estimation, we suggest to estimate the amount of aggregate guarantee. This estimation requires modifications to the fair share estimation function. We refer to the appendix of [8] for informations concerning the implementation of this estimation as well as the deterministic marking.

4.1   Generic Fairness Configuration (GFC) Scenario

The GFC scenario is used to show that flows can benefit from their guarantee and their fair share altogether in networks where the fair share or the RTT varies from one flow to another. This scenario is generic in that the fair share of the links are different and the flows do not have the same path in the network,

---

[5] Some admission control mechanisms that do not require per flow state are proposed in [11].

```
On receiving packet p
    if (edge router)
        i = classify(p);
        AR_i = estimate_rate(AR_i,p);
        /* Determine if the packet is in or out of the guarantees */
        P_out = max(0,1 - (guar_rate_i / AR_i));
        /* Mark the packet */
        if (P_out <= unifrand(0,1))
            p.label = 0;
        else
            p.label = estimate_rate(ER_i,p);
    /* Edge and core routers */
    if (p.label = 0)   /* Packet in the guarantee */
        P_drop = 0;
    else   /* Packet out of the guarantee */
        P_drop = max(0,1-FS/p.label);
    if (P_drop > unifrand(0,1))
        FS = estimate_FS(p,1);
        drop(p);
    else
            if (P_drop > 0)
            p.label = FS;   /* relabel p*/
        FS = estimate_FS(p,0);
        enqueue(p);
```

Fig. 7. Support of minimum guarantees : pseudo-code

leading to different RTTs. Dropping of packets happens in different nodes of the network. Other scenarios and their corresponding simulation results are analysed in [8].

In this scenario, there are 10 sources and 10 destinations (Fig. 8). It follows that 10 flows are considered. These are TCP flows composed of 15 TCP connections each. The flows initiated by the B and X sources are congested on the first link. The flows starting from the C sources are bottlenecked on the second link. And, the flows from the A sources as well as from the Y sources are bottlenecked on the third link. Sources X and Y act as background sources to create congestion at the edge router and at the second core router respectively.

In the Generic Fairness Configuration (GFC) scenario, the first router is an edge router. The second and third routers are classical core routers. They label the packets coming from the sources directly connected to the router. Then, they route the packets to the output links. Finally, the packets are dropped or accepted according to CSFQ, if the router is not connected directly to the packets' destination. The last router is an egress edge router. It does not perform CSFQ. The arriving packets are routed to an output link where they are transmitted to their destination.

The propagation delays on the different links are fixed. The delays of propagation on the links between two routers are 10 ms. Flows A-short, B-short,

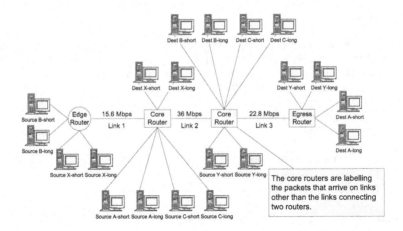

Fig. 8. Generic Fairness Configuration scenario

C-short, X-short and Y-short have a slightly smaller RTT than flows A-long, B-long, C-long, X-long and Y-long. The propagation delay on the links between a source, or destination, of a flow with a small RTT, and a router is 2.5 ms. For flows A-long, B-long, C-long, X-long and Y-long the total delay occurred on the access links has been increased by 32 ms by comparison to the flows with a smaller RTT. This increase is distributed over the two access links crossed by the flows. The window sizes are initialized to 0.6 seconds.

The amount of reserved bandwidth on the second link is 50% of the link rate, in these simulations. Each flow benefits from the same throughput guarantee of 3 Mbps. The rate of link 2 is set to 36 Mbps. 50% of these 36 Mbps are reserved for the TCP flows. There are still 18 Mbps left to share between the A, B and C flows. The excess of the B flows should get 10% of these remaining 18 Mbps. That means that the B flows should be able to use 1.8 Mbps of link 2 in excess of their guarantee. Because there are four flows sharing link 1, the total excess rate on this link should be 1.8 Mbps × 2 = 3.6 Mbps. The rate of link 1 will be set to 4 × 3 Mbps + 3.6 Mbps = 15.6 Mbps. The excess of the B flows should therefore, in average obtain, 1.8 Mbps on link 1. The B flows are not bottlenecked on the second link. It follows that no packet from these flows should be dropped on link 2. A similar reasoning leads to set the rate of the third link to 22.8 Mbps. The deduction is based on the fact that the A flows should have 30% of the bandwidth of link 2 in addition of their guarantees.

4.2 Support of Minimum Guarantees

The goodput of the flows is considered in Fig. 9. It is compared to the goodput that the flows should have in an ideal situation where the guarantees are provided and the excess bandwidth is shared fairly among the flows. The goodput of a flow is the rate of its data transmission. Therefore, the goodput of a flow is smaller than its throughput. In the goodput, packet headers and retransmissions

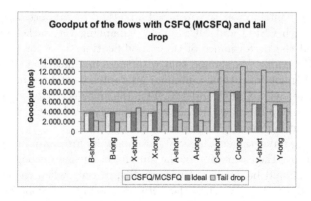

Fig. 9. CSFQ and MCSFQ versus tail drop

of packets are not taken into account. The ideal goodput of Fig. 9 is computed by assuming that no packets are retransmitted which is usually not true. Therefore, it is normal that the goodputs of the flows in the simulations performed with CSFQ (MCSFQ) are slightly below their ideal goodput.

The flows get the same goodput with CSFQ and MCSFQ because the network is always congested and therefore the fair share estimation is the same in both mechanisms. The threshold has also no impact on the bandwidth distribution for the same reason.

When tail drop is used alone, no guarantees are provided to the flows. From Fig. 9, it can be deduced that the A and B flows do not get as much bandwidth as their guarantees. Additionally, the bandwidth is not distributed fairly among the different flows. The RTT has a big impact on the goodput of the flows.

Figure 9 indicates that the flows get their minimum goodput guarantees with CSFQ (MCSFQ). Their excess goodput is above zero. Additionally, the unreserved bandwidth is distributed approximately fairly among the different flows. The flows bottlenecked on the same links approximately get the same amount of bandwidth. The flows with a longer RTT don't get much less bandwidth than the other flows congested on the same link. And, the flows that occur drops at different points in the network, A flows for example, are not too penalized compared to flows that cross less nodes in the network.

5   Conclusion and Further Work

In this paper, we have proposed and evaluated two improvements to CSFQ. First, we have analysed the estimation of the fair share in uncongested networks. We have shown that the estimation used by CSFQ could cause unfairness and underutilization of the network resources. Consequently, we proposed an amelioration to this estimation and validated this improvement by simulations. An important advantage of our Modified CSFQ is that it is much less sensitive to the setting of the window size (K_c) than the original CSFQ.

As a second contribution, we proposed a method to support minimum guaranteed flows with CSFQ and MCSFQ, by enhancing the packet labeling. We have evaluated the performance of these modifications in a complex GFC scenario with TCP traffic. We showed that with these modifications, the TCP flows were able to benefit from their minimum guaranteed bandwidth. Additionally, we noticed that, with CSFQ and MCSFQ, the unreserved bandwidth can be shared fairly among the flows during congestion even when the flows have different RTTs and fair shares.

Several other improvements to CSFQ are possible. First, work should be done to further improve the estimation of the fair share. During uncongested periods, the fair share should be decreased periodically, or only when the congestion is noticed, to avoid tail drops when the network starts to be congested, because tail drops lead to unfair bandwidth distribution. During congestion, another estimation method could also be used. The applicability of the estimation used in the Fair Allocation Derivative Estimation (FADE) [5] should be studied. Second, when tail drops occur with CSFQ, the fair share is decreased by a small percentage [10]. It could be interesting to avoid tail drops instead of reacting to tail drops. The mechanism proposed in [4] could be used for this purpose [8].

Acknowledgments

We would like to thank Olivier Bonaventure for his constructive comments as well as Steve Uhlig, Pierre Reinbold and Bruno Quoitin for their reviews.

References

[1] S. Blake, D. Black, M. Carlson, E. Davies, Z. Wang, and W. Weiss. An architecture for differentiated services. Internet RFC 2475, December 1998.

[2] R. Braden, D. Clark, and S. Shenker. Integrated services in the Internet architecture : an overview. Internet RFC 1633, July 1994.

[3] Z. Cao, Z. Wang, and E. Zegura. Rainbow fair queueing: Fair bandwidth sharing without per-flow state. In *Proceedings INFOCOM '00*, March 2000.

[4] Stefaan De Cnodder, Kenny Pauwels, and Omar Elloumi. A rate based RED mechanism. In *Proc. of the 10th International Workshop on Network and Operating System Support for Digital Audio and Video*, NOSSDAV 2000, Chapel Hill, NC, 26–28 June 2000.

[5] Na Li, Marissa Borrego, and San-qi Li. Achieving per-flow fair rate allocation within diffserv. In *Proceedings of the fifth IEEE Symposium on Computers and Communications*, ISCC 2000, Antibes, France, 3-6 July 2000.

[6] M. Nabeshima, T. Shimizu, and I. Yamasaki. Fair queueing with in/out bit in core stateless networks. In *Proc. of the 8th International Workshop on Quality of Service*, IWQoS 2000, Pittsburgh, PA, 5–7 June 2000.

[7] Kenny Pauwels, Stefaan De Cnodder, and Omar Elloumi. A multi-color marking scheme to achieve fair bandwidth allocation. In *Proc. of the 1st International Workshop on Quality of future Internet Services*, QofIS 2000, Berlin, Germany, 25–26 September 2000.

[8] Cristel Pelsser. Support of fairness and guarantees without per-flow state in routers. Master's thesis, Facultés Universitaires Notre-Dame de la Paix, Namur, Belgium, June 2001. Available at `www.info.fundp.ac.be/\char126cpe/memoire.ps.gz`.

[9] I. Stoica, H. Zhang, S. Shenker, R. Yavatkar, D. Stephens, A. Malis, Y. Bernet, Z. Wang, F. Baker, J. Wroclawski, C. Song, and R. Wilder. Per hop behaviors based on dynamic packet states. Internet Engineering Task Force, Work in Progress, draft-stoica-diffserv-dps-00.txt, February 1999.

[10] Ion Stoica, Scott Shenker, and Hui Zhang. Core-stateless fair queueing : Achieving approximately fair bandwidth allocations in high speed networks. In *Proceedings of ACM SIGCOMM'98*, Vancouver, BC, October 1998.

[11] Ion Stoica and Hui Zhang. Providing guaranteed services without per flow management. In *Proceedings of ACM SIGCOMM'99*, Cambridge, MA, August 1999.

[12] N. Venkitaraman, J. Mysore, R. Srikant, and R. Barnes. Stateless prioritized fair queueing. Internet Engineering Task Force, Work in Progress, draft-venkitaraman-spfq-00.txt, July 2000.

A Fast Packet Classification
by Using Enhanced Tuple Pruning

Pi-Chung Wang[1], Chia-Tai Chan[1], Wei-Chun Tseng[1], and Yaw-Chung Chen[2]

[1] Telecommunication Laboratories, Chunghwa Telecom Co., Ltd,
7F, No. 9 Lane 74 Hsin-Yi Rd. Sec. 4, Taipei, 106 Taiwan, R.O.C.,
Tel: +886-2-23265631, Fax: +886-2-23445700,
{abu,ctchan,wctseng}@cht.com.tw
[2] Department of Computer Science and Information Engineering,
National Chiao Tung University, Hsinchu, Taiwan, R.O.C.,
{ycchen}@csie.nctu.edu.tw

Abstract. In the packet classification, the route and resources allocated to a packet are determined by the destination address as well as other header fields of the packet such as source/destination address, TCP and UDP port numbers. It has been demonstrated that performing packet classification on a potentially large number of fields is difficult and has poor worst-case performance. In this work, we proposed an enhanced tuple pruning search algorithm called *"Tuple Pruning +"* that provides fast two-dimension packet classification. With reasonable extra filters added for *Information Marker*, only one hash access to the tuples is required. Through experiments, about 8 MB memory is required for 100K-filter database and 20 million packet per second (MPPS) is achievable. The results demonstrate that the proposed algorithm is suitable for high-speed packet classification.

1 Introduction

Traditionally, routers have forwarded packets based only on the destination address of the packet and do not provide service differentiation because they treat all traffic going in the same way. Increasingly, new services require more discriminating forwarding, called "Packet Classification". It allows service differentiation because the router can distinguish traffic based on source/destination address and application type. The process of mapping packets to different service classes is referred to as packet classification. The simplest, best-know form of packet classification is IP lookups, in which each rule specifies a destination prefix. The associated action is the IP address of next router that the packet must be forwarded. The other services which require packet classification include: access-control of firewalls, policy based routing, provision of differentiated qualities of services, and traffic billing, etc.

To describe the problem formally, we have to define the classifier and the filter. A classifier is a set of rules or filters that specifies the flows or classes. Packet classification is performed using a packet classifier, also called filter database. A

G. Carle and M. Zitterbart (Eds.): PfHSN 2002, LNCS 2334, pp. 180–191, 2002.

filter F is called k tuple $F = (f[1], f[2], \ldots f[k])$ if the filter contents the k fields of the packet header, where each $f[i]$ is either a variable length prefix bit string, a range or a explicit value. A filter can be combined from many fields, for a packet header, the most common fields are the IP source address (SA), the destination address (DA), the protocol type and port numbers of source and destination applications and protocol flags. A packet P is said to match a particular filter F if for all i, the i_{th} field of the header satisfies the $f[i]$. Each filter has an associative action. For example, the filter F=(140.113.*, *, tcp, 23, *) specifies a rule that flows which address to the subnet 140.113 use the telnet application and the action of the rule may disallow these flows into its network. Beside of the action, the filter is usually given a cost value to define the priority in the database. The action of the least-cost matching filter will be used to process the arriving packet.

To perform packet classification on a potentially large number of filters on key header fields is difficult and has poor worst-case performance. In the previous work, tuple pruning search is proposed to achieve fast and scalable two-dimension (SA,DA) packet classification [1]. Through simulation, 11 hash accesses are required to finish a packet classification in the worst case as described in the literature. In this work, an enhanced tuple pruning search algorithm is proposed. With reasonable extra filters added, only one hash access to the tuples is required. Also, about 8 MB memory is required for 100,000-filter database. By using parallel hardware design, 20 million packet per second (MPPS) is achievable. The proposed algorithm is thus suitable for high speed packet classification.

The rest of the paper is organized as follows. Firstly, the related algorithms are introduced in Section 2. Section 3 presents the proposed algorithm. The experiment setup and results are presented in 4. Finally, a summary is given in Section 5.

2   Related Works

Recently several algorithms for packet classification have appeared in the literature [1], [2], [3], [4], [5], [6], [7], [8], [9]. It can be categorized into following classes: linear search/caching, hardware-based, grid of tries/cross-producting, recursive-flow classification, hierarchical intelligent cuttings, and hash-based solution. Many of these algorithms which provide fast lookup performance, required $O(N^k)$ memory space in the worst case, where N is number of filters and k is the number of classified fields. In the following, we briefly described the main properties of these algorithms.

Linear Search/Caching: The simplest approach to packet classification is to perform a linear search through all the filters. This requires $O(N)$ memory, but also takes $O(N)$ lookup time, which would be unacceptably large even for modest size filter sets. Caching is a technique often employed at either hardware or software level to improve performance of linear search. If packets from the same flow have identical headers, packet headers and corresponding classification solution can be cached. However, performance of caching is critically dependent on having large number of packets in each flow. Also, if number of simultaneous flows becomes larger than cache size, performance degrades severely. Note that the average lookup time is adversely affected by even a small miss rate due to very high cost of linear search. Hence caching is much more useful when combined with a good classification algorithm that has a low miss penalty.

Hardware-Based Solutions: A high degree of parallelism can be implemented in hardware to gain speed-up advantage. Particularly, Ternary Content Addressable Memories (TCAM) can be used effectively for filter lookup. However,TCAM with particular word width cannot be used when flexibility in filter specification to accommodate larger filters is desired. It is difficult to manufacture TCAM with wide enough words to contain all bits in a filter. An interesting approach that relies on very wide memory bus is presented by Lakshamn et $al.$ [4]. The scheme computes the best matching prefix for each of the k fields of the filter set. For each filter a pre-computed N-bit bitmap is maintained. The algorithm reads Nk bits from memory, corresponding to the best matching prefixes in each field and takes their intersection to find the set of matching filters. Memory requirement for this scheme is $O(N^2)$ and it requires reading Nk bits from memory. These hardware-oriented schemes rely on heavy parallelism, and requires significant hardware cost, not to mention that flexibility and scalability of hardware solutions is very limited.

Grid of Tries/Cross-Producting: For the case of 2-field filters, Srinivasan et $al.$ presented a trie-based algorithm [3]. This algorithm has memory requirement $O(NW)$ and requires $2W - 1$ memory accesses per filter lookup. A general mechanism called cross-producting is also presented. It involves performing best matching prefix lookups on individual fields, and using a pre-computed table for combining results of individual prefix lookups. However, this scheme suffers from a $O(N^k)$ memory blowup for k-field filters, including $k = 2$ field filters.

Recursive-Flow Classification: Gupta et $al.$ presented an algorithm, which can be considered as a generalization of cross-producting [4]. After best matching prefix lookup has been performed, recursive flow classification algorithm performs cross-producting in a hierarchical manner. Thus k best matching prefix lookups and $k - 1$ additional memory accesses are required per filter lookup. It is expected to provide significant improvement on an average, but it requires $O(N^k)$ memory in the worst case. Also, for the case of 2-field filters, this scheme is the same as cross-producting and hence has memory requirement of $O(N^2)$.

Hierarchical Intelligent Cuttings: Gupta et $al.$ proposed a heuristic HI-Cuts that makes hierarchical cuts in the search space [8]. It is difficult to characterize conditions under which such heuristics perform well, and the worst-case memory utilization for the HICuts scheme may explode.

3   Enhanced Tuple Pruning

3.1   Tuple Space Search

The tuple space idea generalizes the aforementioned approach to multi-dimensional filters [10]. A tuple T is defined as a combination of field length, and the resulting set is called tuple space. Since each tuple has a known set of bits in each field, by concatenating these bits in order we can create a hash key, which can then be used to map filters of that tuple into a hash table. As an example, the two-dimensional filters $F=(10^*, 110^*)$ and $G=(11^*, 001^*)$ will both map to $T_{2,3}$. When searching $T_{2,3}$, a hash key is constructed by concatenating 2 bits of the source field with 3 bits of the destination field. Thus, the best matching filter can be found by probing each tuple alternately, and keeping track of the best matching filter. Since the number of tuples is generally much smaller than the number of filters, even a linear search of the tuple space results in a significant improvement over linear search of the filters.

To improve the speed of linear search, pre-computation and marker is used [1]. As a result, $2W - 1$ hash probes are required where W is the length of IP address. Another heuristic, tuple space pruning, performs lookups on individual fields to eliminate tuples that cannot match the query. Although this heuristic does not provide any improvement in the worst case, it performs well in the practical environment. Through the experiment, the number of probed tuples is reduced to about 10 in the worst case.

3.2 Enhance Mechanism: Tuple Pruning +

The tuple pruning search can be improved by adopting the concept of best match prefix (BMP). In the BMP problem, the longest matched prefix in the lookup procedure will be chosen to identify the route. Since there is only one longest matched prefix for each IP address, we can assign the filters related to the IP address to the tuple according to the longest matched prefix. Thus only the tuple with the BMP needs to be probed in the packet classification. Assume there are two filters in the two-dimension classifier: $(10^*,110^*)$ and $(1010^*,110010^*)$, as shown in Figure 1. These two filters will be assigned to the tuples according to their longest matched prefix, thus they will be located to $T_{2,3}$ and $T_{4,6}$, respectively. To further improve the tuple pruning search, an information marker is introduced to maintain the associated information for future tuple search.

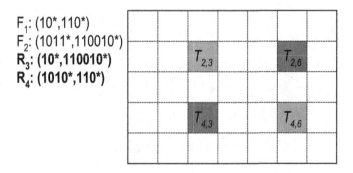

Fig. 1. A Sample Classifier with Two Filters.

By applying the idea in our proposed mechanism, one extra entry R_3 will be generated at $T_{2,6}$ and its action is equal to that of F_1. This filter will be referred for packets which SA is matched to 10* and DA is matched to 110010*, for example, the packet with header (100000,110010). According to the proposed idea, the lookup procedure will refer tuple (2,6), thus the associated action will be selected. We name this filter as a information-marker (i-marker here after). It is used to improve the search procedure as the marker used in [1], the major different is that i-marker is only put to the tuples with longer prefixes. The i-marker should also be added to $T_{4,3}$ and $T_{4,6}$. However, the one inserted into $T_{4,6}$ is identical to F_2, thus the action of F_1 will be compared with that of F_2. If the cost of F_1 action is lower, its action will occupy the action of F_2.

One of the major concerns about this approach is the number of the additional i-markers. Apparently, the number of the i-markers ties to the number of tuples with shorter prefix for each IP address. To illustrate this problem, we use the routing tables downloaded from [11], [12] as an example. In Figure 2, we show the number of shorter prefixes for each route prefix without counting the default route. Obviously, for most route prefixes, there are usually less than three shorter prefixes in the routing table and six in the worst case. On the other hand, at most 48 ($7^2 - 1$) extra filters might be generated for each inserted filter. However, the occurrence of the worst-case situation should be relatively low since only 5% of route prefixes have more than three and two shorter prefixes in the NLANR and the rest routing tables, respectively. And also, each shorter prefix may not appear in the classifiers. As a result, we believe that the extra cost should be acceptable with respect to the performance improvement.

Tuple Construction: To build the classification tuples, the procedure consists of two parts. First, each filter is inserted into the associated tuple according to its length. In the mean time, a prefix tree should be constructed to record the referred prefixes in the filters. A binary tree or the multi-bit tree proposed in

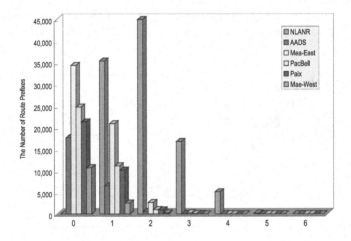

Fig. 2. The Number of Shorter Prefixes for Each Route Prefix.

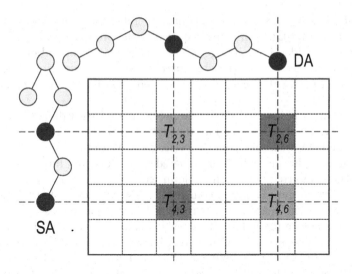

Fig. 3. The Tuple Construction from the Prefix Tree.

[13] can be used as the prefix tree. Then, the pruning table for each dimension will be generated from the prefix tree. Also, it will keep track of the relationship between each route prefix and its shorter prefixes, as shown in Figure 3. Second, the i-marker is added to the tuples from the prefix tree. For example, if there are m longer prefixes for a given SA and n longer prefixes for a given DA, then there are at most $m \times n$ i-markers will be added to the tuples. Before the i-marker is inserted into the tuple, the existence of duplicate filter is checked. If there is no duplicate entry, the i-marker is inserted. Otherwise, the cost of filter actions will be compared for the tuples with duplicate filter and the lower-cost action will be recorded in the entry. For the sake of ease update, each entry in the tuple should have two action fields: one is the lowest-cost action related to the filter and another is its original action.

Note that the insertion of i-markers will not affect the construction of pruning table since it is based on the original filters. Furthermore, the filters with at least one wildcard will not be inserted into the tuples. Those filters only need to be inserted into the prefix tree since they can be treated as single dimension prefixes. Thus it can reduce the number of i-markers.

Search: The classification procedure consists of two lookups of pruning tables and one hash lookup to the tuple. Firstly, the BMP lookup is performed in pruning tables for each dimension. However, the lookup result fetched here is different from that of IP route lookup, in which only the length of the BMP is needed in the pruning lookup. After the length of two BMPs l_1 and l_2 are found, the tuple (l_1, l_2) will be probed for the matched filter. Obviously, the tuple space lookup performance mainly ties to the lookup performance of pruning tables. The fast lookup algorithm proposed in the previous schemes can be applied to provide good performance.

Update: The update procedure is a little complex due to the pre-computation. However, the re-construction is not required in the proposed scheme. The table update can be divided into: change of filter action, insertion and deletion of filter. We only explain how to perform filter insertion and deletion. The change of filter action is trivial since it can be treated as re-insert the filter with new action.

To deal with the inserted filter, the prefix tree for each dimension is maintained. Firstly, it will be inserted into the prefix tree, and then the longer prefixes for the SA and DA are found from the prefix trees. With the lengths of the longer prefixes, the set of tuples which are covered by the inserted filter is calculated. An i-marker is inserted into each tuple in the set. If there is a filter with the same key as the i-marker, a cost comparison is performed and the action with lower cost will be left in the entry. Furthermore, the tuples covered by the lower-cost filter will not be probed for the insertion since they will not be affected by the inserted filter, as shown in Figure 4. A filter F_5 is inserted into the two-filter database of Figure 1. After inserting the SA and DA into the prefix trees, the set of probed tuples are derived. According to the row-major order, the i-markers are put into $T_{1,3}$, $T_{1,6}$, $T_{2,2}$ and $T_{4,2}$, respectively. While traversing $T_{2,3}$, a collision with F_1 is encountered. After comparing their cost, if the cost of F_5 is lower, its action will replace the lowest-cost action field of F_1 and keep traversing the remaining tuples. Otherwise, the entry in $T_{2,3}$ will remain unchanged and the remained three tuples ($T_{2,6}$, $T_{4,3}$ and $T_{4,6}$) which are covered by F_1 will not be probed in this insertion. In the worst case, it will update W^2 tuples.

The procedure of filter deletion is a little similar to that of filter insertion. Now we use the filter database with newly inserted F_5 in Figure 5 as an example. If the filter F_1 is deleted from the database, the tuples covered by it will be probed for possible update. However, before the tuple-probe proceeding, the nearest filter which covers F_5 should be found for possible referring. This is because if there are probed tuples with i-markers or filters with cost higher than F_1 and F_5, the action of F_5 should replace those entries to ensure that the lowest cost

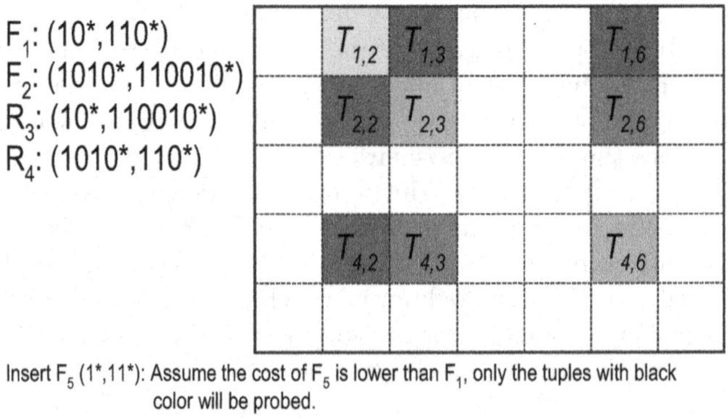

Insert F_5 (1*,11*): Assume the cost of F_5 is lower than F_1, only the tuples with black color will be probed.

Fig. 4. An Example of Filter Insertion.

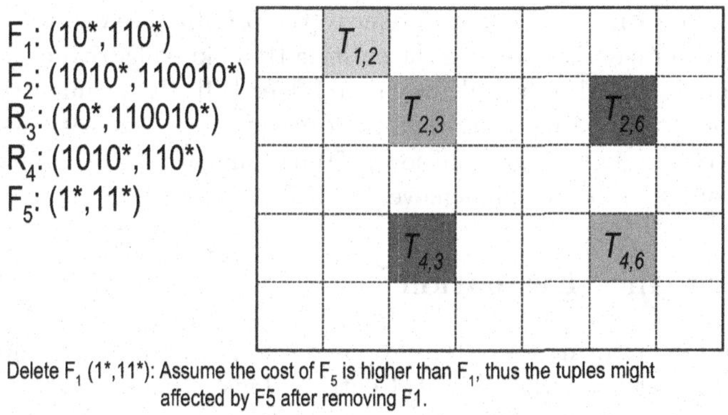

F_1: (10*,110*)
F_2: (1010*,110010*)
R_3: (10*,110010*)
R_4: (1010*,110*)
F_5: (1*,11*)

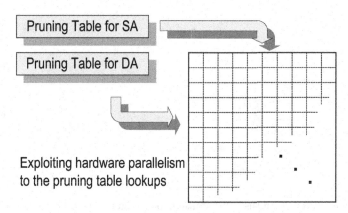

Delete F_1 (1*,11*): Assume the cost of F_5 is higher than F_1, thus the tuples might affected by F5 after removing F1.

Fig. 5. An Example of Filter Deletion.

action will be taken. Consequently, the i-markers in the probed tuples will be refreshed by F_5. Also, the filters will be checked whether the cost is higher than F_5. If yes, they will be occupied by the action of F_5. Otherwise, their actions will be used instead. The time complexity for update is also $O(W^2)$.

Implementation: The pruning tuple space search can be implemented with software or hardware. With software implementation, the total lookup time is $(lookup(SA) + look(DA))$ plus one hash access time. To deal with the potential large number of entries, the hash function can allocate multiple entries in a pool to fit the cache line.

The lookup performance can be further improved through hardware implementation. By exploiting hardware parallelism, the total lookup time of the pruning tables is reduced to $max(lookup(SA), lookup(DA))$, as shown in Figure 6. It can also perform the pruning and hash simultaneously by adopting pipeline design to achieve higher throughput. As a result, we can accomplish one packet classification within *maximum(pruning(SA), pruning(DA), one hash access to the tuple)*.

Pruning Table for SA

Pruning Table for DA

Exploiting hardware parallelism
to the pruning table lookups

Fig. 6. Implement with Parallel Hardware.

Since the most memory accesses refer to the pruning, we can use high-speed SRAM as its storage. By adopting the existing IP lookup algorithms for pruning, such as multibit trie [13] or multiway search tree [14], the pruning time can be reduced to less than 50 ns easily (less than ten memory accesses). Assume that one hash access time without collision is 50 ns (one 50-ns DRAM access time), thus the proposed scheme can achieve 20 MPPS.

4 Performance Evaluation

To evaluate the performance of the proposed scheme, we use the randomly generated filter database with 5K to 1M entries. This is mainly because that the filter database is usually considered as secret data commercially. Also, most of them are relatively small, such as the filter databases used in [4]. Thus we generate the filter database from the routing table in NLANR. There are 102,309 prefixes in the sample routing table [11]. We use two different sampling schemes to generate the (SA,DA) filters: the first one is to choose the prefixes uniformly [1] and the other is to concentrate 80% filters in 20% address space to show the locality [9]. Note that the filters with wildcard are not considered in the simulation because they will be inserted into the pruning table and will not affect the tuples. The filter length distribution with 100,000 filters with 80% locality is shown in the right part of Figure 7 which is similar to the figure of the uniformly chosen filters. Obviously, most filters correspond to the tuples near (24,24). A darker color indicates that there are more filters in the tuple.

We first examine the filters database with 80% locality. The major performance metrics ties to the number of i-markers. Since the size of i-marker is equal to the filter, we use the term "entry" to cover both. From Table 1, we can see that the numbers of entries are about three to six times of the original tables. However, with a larger database (larger than 10,000 entries), the increased entry ratio is lower with respect to the smaller database (1,000). This is because with more entries in the table, the probability to generate an i-marker with collision filter is

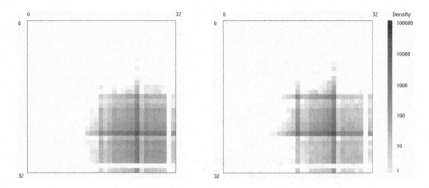

Fig. 7. The Filter Lengths Distribution of Original Database. (Left:Random, Right:80% Locality)

Table 1. The Number of Entries versus the Pruned Tuples without *i*-Marker. (80% Locality)

Filter	Original Scheme			Proposed Scheme		
Count	Tuples	Probes	Entry Count	Tuples	Probes	Entry Count
1,000	129	8	1,000	153	1	5,623
5,000	224	11	5,000	249	1	14,105
10,000	295	11	10,000	318	1	30,654
50,000	365	15	50,000	377	1	162,931
100,000	353	16	100,000	360	1	293,217
500,000	442	30	500,000	462	1	2,400,103
1,000,000	504	51	1,000,000	530	1	5,628,952

also higher, this reduces the ratio of increased entries. It does suggest good scalability, especially under the speed-critical environment, the proposed scheme has apparent improvement. The result for the random-generated database is shown in Table 2. The number of pruned tuple is slightly reduced because the address locality may result in more intersection in the pruning. However, the number of entries is increased for the large database (for database with more than 50,000 filters); this is because that large amount filters result in more related prefixes in the database. However, without *i*-marker, the probed tuples will increased to 51 in the worst case for 1M-filter database, i.e., at least 51 memory accesses. Obviously, the speed improvement is necessary, even with about 6 times storage.

For the random-generated database, the number of generated entries is more than that in the 80%-locality database. This is because the wide-spread filters might cause more *i*-markers in the worst case. We assume that the memory utilization of the hash table is 50%. Thus, for the 100k-filter database, it requires about 8 MB memory, whose cost is lower than US$ 10. However, it can achieve about 20 MPPS by using the 50-ns DRAM. While the database enlarges to 1M filter, it will requires about 130 MB memory without speed degradation.

Table 2. The Number of Entries versus the Pruned Tuples without *i*-Marker. (Random)

Filter	Original Scheme			Proposed Scheme		
Count	Tuples	Probes	Entry Count	Tuples	Probes	Entry Count
1,000	139	2	1,000	140	1	2,009
5,000	242	4	5,000	246	1	12,438
10,000	274	5	10,000	276	1	23,583
50,000	334	12	50,000	341	1	195,990
100,000	361	14	100,000	375	1	374,718
500,000	440	31	500,000	459	1	2,685,592
1,000,000	468	41	1,000,000	491	1	6,814,934

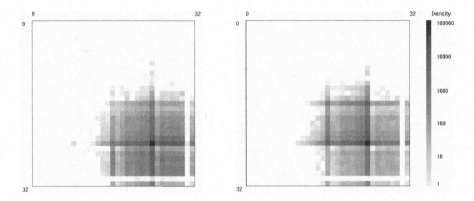

Fig. 8. The Filter Lengths Distribution of Database with *i*-Markers. (Left:Random, Right:80% Locality)

The filter lengths distribution graph from Figure 7 is shown in Figure 8. One can see that the number of required tuples is increased and the colors of most of the blocks are darker than that in the previous graph. Furthermore, the number of colored blocks is increased because the *i*-markers might be inserted to the tuples without filter originally.

5   Conclusion

In this paper, we propose a remarkable enhancement to the previous work. By using the pre-computation and *i*-markers, we can reduce the number of probed tuples from the worst-case $O(W^2)$ to $O(1)$. The incremental update is also supported. In the worst case, the number of generated *i*-markers are four times of the original filters for the 100,000-filter datebase. From the simulation, the proposed scheme with parallel design can achieve 20 MPPS in the worst case. In the future, we will focus on dynamic space-cutting algorithm to further reduce the required storage.

References

1. V. Srinivasan, G. Varghese and S. Suri: Packet Classification using Tuple Space Search. ACM SIGCOMM. (1999) 135–146
2. T.V. Lakshman and D. Stidialis: High Speed Policy-based Packet Forwarding Using Efficient Multi-dimensional Range Matching. ACM SIGCOMM. (1999) 203–214
3. V. Srinivasan, G. Varghese, S. Suri and M. Waldvogel: Fast Scalable Level Four Switching. ACM SIGCOMM. (1998) 191–202
4. Pankaj Gupta and Nick McKeown: Packet Classification on Multiple Fields. ACM SIGCOMM. (1999) 147–160
5. Anja Feldmann and S. Muthukrishnan: Tradeoffs for Packet Classification. IEEE INFOCOM. (2000) 1193–1202

6. Thomas Woo: A Modular Approach to Packet Classification: Algorithms and Results. IEEE INFOCOM. (2000) 1213–1222
7. M. Buddhikot, S. Suri and M. Waldvogel: Space Decomposition Techniques for Fast Layer-4 Switching. IFIP Sixth International Workshop on High Speed Networks. (2000)
8. Pankaj Gupta and Nick McKeown: Packet Classification using Hierarchical Intelligent Cuttings. Hot Interconnects VII. (1999)
9. Ying-Dar Lin, Huan-Yun Wei and Kuo-Jui Wu: Ordered lookup with bypass matching for scalable per-flow classification in layer 4 routers. Computer Communications, Vol. 24. (2001) 667–676
10. M. Waldvogel, G. Varghese, J. Turner and B. Plattner: Scalable High Speed IP Routing Lookups. ACM SIGCOMM. (1997) 25–36
11. NLANR Project: National Laboratory for Applied Network Research. See http://www.nlanr.net
12. Merit Networks Inc.: Internet Performance Measurement and Analysis (IPMA) Statistics and Daily Reports. IMPA Project. See http://www.merit.edu/ipma/routing_table/
13. V. Srinivasan and G. Varghese: Fast IP lookups using controlled prefix expansion. ACM Trans. On Computers, Vol. 17. (1999) 1–40
14. P.C. Wang, C.T. Chan and Y.C. Chen: Performance Enhancement of IP forwarding by using Routing Interval. Journal of Communications and Networks, Vol. 3. (2001) 374–382.

Traffic Engineering with AIMD in MPLS Networks*

Jianping Wang[1], Stephen Patek[2], Haiyong Wang[1], and Jörg Liebeherr[1]

[1] Department of Computer Science
[2] Department of Systems and Information Engineering,
University of Virginia, Charlottesville, VA 22904, U.S.A.,
{jgwang@cs,hw6h@cs,jorg@cs,patek}@virginia.edu

Abstract. We consider the problem of allocating bandwidth to competing flows in an MPLS network, subject to constraints on fairness, efficiency, and administrative complexity. The aggregate traffic between a source and a destination, called a flow, is mapped to label switched paths (LSPs) across the network. Each flow is assigned a preferred ('primary') LSP, but traffic may be sent to other ('secondary') LSPs. Within this context, we define objectives for traffic engineering, such as fairness, efficiency, and preferred flow assignment to the primary LSP of a flow ('Primary Path First', PPF). We propose a distributed, feedback-based multipath routing algorithm that attempts to apply additive-increase and multiplicative-decrease (AIMD) to implement our traffic engineering objectives. The new algorithm is referred to as multipath-AIMD. We use *ns-2* simulations to illustrate the fairness criteria and PPF property of our multipath-AIMD scheme in an MPLS network.

1 Introduction

Multiprotocol Label Switching (MPLS) [20] has offered new opportunities for improving Internet services through traffic engineering. An important aspect of traffic engineering, defined as "that aspect of Internet network engineering dealing with the issue of performance evaluation and performance optimization of operational IP networks" [4], is the allocation of network resources to satisfy an aggregate measure of the demand for services and to obtain better network utilization. MPLS, in conjunction with path establishment protocols such as CR-LDP [2] or RSVP-TE [8], makes it possible for network engineers to set up dedicated label switched paths (LSPs) with reserved bandwidth for the purpose of optimally distributing traffic across a given network.

Figure 1 illustrates an MPLS network, where all traffic across the network is accounted for by a set of source/destination pairs, called flows, and multiple LSPs are in place for accommodating the demand for service. We consider a multipath routing scenario where sources can make use of multiple LSPs. For each source, one LSP is

---

* This work is supported in part by the National Science Foundation through grants ANI-9730103, ECS-9875688 (CAREER), ANI-9903001, DMS-9971493, and ANI-0085955.

G. Carle and M. Zitterbart (Eds.): PfHSN 2002, LNCS 2334, pp. 192–210, 2002.
© Springer-Verlag Berlin Heidelberg 2002

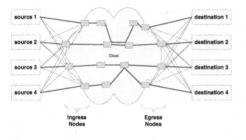
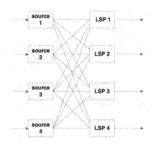
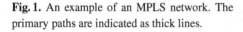

Fig. 1. An example of an MPLS network. The primary paths are indicated as thick lines.

Fig. 2. The simplified network model. LSP i is the primary path for source i.

assigned as the primary path, and other LSPs can be used as secondary paths. We consider sets of sources where each source can use all primary paths of the other sources as its secondary paths. While Figure 1 presents a general view of an MPLS network, we concentrate in this paper on a simplified model, as illustrated in Figure 2. To begin, we assume that there are N sources and N LSPs, each serving as the primary path associated with exactly one source. In this context, the traffic engineering problem is the assignment of traffic of a flow to the primary path and the secondary paths, in such a way that a given set of traffic engineering objectives is satisfied. While a centralized solution to the given problem is quite straightforward, we strive to find distributed mechanisms for traffic engineering without central control. Specifically, we investigate if and to which degree binary feedback schemes and rate control schemes, such as, additive increase/multiplicative decrease (AIMD) [5,6,10,11,12,19], can be used to achieve traffic engineering objectives.

Recently, considerable effort has been invested into scalable mechanisms for providing differentiated services in the Internet. For example, in [7], Elwalid et al. presented a multi-path adaptive traffic engineering mechanism, called MATE, designed for MPLS networks where several explicit LSPs have been established between ingress and egress nodes. MATE is intended to work for traffic that does not require bandwidth reservation and seeks to distribute traffic across the LSPs by dynamically adjusting the rate on each ingress node. The relationships between end-to-end congestion control and fairness are established in [9,14,15,17,18,21]. The network model adopted in these papers supposes that all sources are greedy and each source sends traffic through a single path or a dedicated set of paths, and fairness is characterized by means of a social welfare-type optimization objective. These models generally give rise to differential equations that characterize the behavior of AIMD and AIPD[1] congestion control schemes.

Different from most of the related work on fairness and binary feedback, we consider sources as being either greedy or non-greedy. A source is greedy if it always has traffic backlogged. A non-greedy source has an upper bound on the desired sending rate.

---
[1] additive-increase/proportional-decrease

Moreover, we model LSPs as being either pooled or owned. In the owned case, each LSP gives high priority to the source for which the path is primary, and any remaining capacity is available to be shared equally among the remaining sources. In the pooled case, primary paths do not give priority to their respective sources.

In this paper, we propose congestion control mechanisms for dynamically adjusting the rates of all sources. In contrast to [9,14,15,17,18,21], in an effort to minimize overhead and processing complexity, we only employ binary feedback mechanisms in adjusting flow rates. The feedback mechanisms we have developed seek to address the following traffic engineering characteristics.

• **Efficiency.** An allocation of network resources is said to be efficient if either all resources are completely consumed while the network is overloaded or all sources are completely satisfied while the network is underloaded.

• **Fairness.** The appropriate notion of fairness for MPLS traffic engineering varies with the nature of sources (greedy vs. non-greedy) and LSPs (pooled vs. owned). In Section 2, we formally define the corresponding fairness criteria for each case, based on the notion of fair-share resource allocation in [3]. With N greedy sources and N pooled resources, it is easy to prove the equivalence between our definition and the minimum potential delay fairness from [16,18]. However, our fairness definitions are constructive, providing an easier way to characterize and achieve fair allocations than through the solution of nonlinear optimization models.

• **Primary Path First Property.** While routing along multiple paths is an opportunity we seek to exploit, there are drawbacks associated with multi-path routing, such as overhead associated with label distribution, additional state information, classification, and potential out-of-sequence delivery. To address these issues, we formulate a traffic engineering objective that seeks to minimize the amount of traffic sent over secondary paths. We introduce a novel criteria for network performance, called the Primary Path First (PPF) property. Generally speaking, PPF refers to the desire to have each source exploit available capacity on secondary paths, but to refrain from using the secondary paths whenever possible. We make this notion precise in Section 3. The PPF property reflects that, given multiple feasible rate assignments to primary and secondary paths that satisfy fairness and efficiency criteria, the preferred assignment is the one that sends the most traffic of a source on the primary path of that source.

• **Distributed Implementation.** We seek to provide efficient, fair, and PPF allocations of network resources using simple distributed algorithms. Distributed mechanisms, which operate mainly on local state information, are preferred as they minimize network overhead and retain scalability. In Section 4, we describe a distributed scheme which allocates flow to resources in a fashion reminiscent of AIMD in [13]. In some cases, global information and coordination are required for specific goals such as PPF.

• **Stability and Convergence.** Traffic engineering mechanisms, such as those we seek to develop in this paper, often suffer from potential instability and oscillations within the network. We seek to prevent this type of behavior by requiring incremental adjustments to the flow allocations specified by our algorithms. However, in this paper we do not offer a formal proof of stability or convergence properties.

This paper makes two contributions. First, we introduce notions of fairness for MPLS traffic engineering, and we show how the AIMD algorithm of [13] can be extended, in a distributed fashion, to achieve fair allocations of network resources. We call the enhanced AIMD algorithm multipath-AIMD. Next, we introduce the PPF criterion which seeks to limit the administrative complexity associated with multipath-routing, by concentrating traffic on a designated LSP.

The reminder of this paper is organized as follows. Section 2 formally defines notions of fairness for both pooled and owned resources. Section 3 defines and analyzes the PPF criterion. Section 4 presents AIMD algorithms which experimentally converge to fair allocations of network capacity and suggests modifications to the AIMD schemes for achieving a PPF assignment of resources. Section 5 presents *ns-2* simulation results, and Section 6 concludes the paper.

2   Fairness Criteria

We consider a network of LSPs which correspond to the simplified model of Figure 2. In this model, there are N traffic sources and N LSPs.

At any time, each source $i = 1, \ldots, N$ has a load of λ_i ($\lambda_i \geq 0$) , which is the maximum desired sending rate of the source. If the traffic demand from source i is $\lambda_i = \infty$, we say that the source is *greedy*.[2] Each LSP $i = 1, \ldots, N$ has a maximum transmission capacity of B_i. LSP i is the primary path associated with source i, and all other LSPs are secondary paths with respect to source i. We use γ_i to denote the actual allocation of bandwidth to source i. The rate allocation consists of the allocation on the primary path and the secondary paths.

We distinguish two different allocation schemes for assigning bandwidth on the LSPs to sources:

- **Owned Resources:** Each source may consume the entire capacity of its primary path, i.e., B_i, and, in addition, it can obtain unused bandwidth on its secondary paths.
- **Pooled Resources:** The aggregate capacity on all LSPs, i.e., $\sum_{i=1}^{N} B_i$, is distributed across all sources, without regard to the capacity on primary paths.

The fairness criteria for networks with owned and pooled resources are specified in the following definition.

Definition 1. *Given a network as shown in Figure 2 with N sources and LSPs. Let B_i denote the capacity of LSP i ($1 \leq i \leq N$), and let $\lambda_i \geq 0$ and $\gamma_i \geq 0$ denote the load and the rate allocation of source i.*

---

[2] Note that the values of λ_i vary with time; However, we do not carry the dependence on time in our notation, i.e., by writing '$\lambda_i(t)$'.

1. *A rate allocation is a relation* $R = (\lambda_i, \gamma_i)$, $1 \leq i \leq N$ *such that both* $\gamma_i \leq \lambda_i$ *and*
   $0 \leq \sum_{i=1}^{N} \gamma_i \leq \sum_{i=1}^{N} B_i$.
2. *A rate allocation is efficient if the following hold:*
   a) If $\sum_{i=1}^{N} \lambda_i < \sum_{i=1}^{N} B_i$, *then* $\sum_{i=1}^{N} \gamma_i = \sum_{i=1}^{N} \lambda_i$,
   b) If $\sum_{i=1}^{N} \lambda_i \geq \sum_{i=1}^{N} B_i$, *then* $\sum_{i=1}^{N} \gamma_i = \sum_{i=1}^{N} B_i$.
   If case b) holds, we say that the rate allocation is saturating.
3. *A rate allocation for pooled resources is fair if there exists a value* $\alpha^p > 0$ *such that for each source* i *it holds that* $\gamma_i = \min\{\lambda_i, \ \alpha^p\}$.
4. *A rate allocation for owned resources is fair if there exists a value* $\alpha^o > 0$ *such that for each source* i *it holds that* $\gamma_i = \min\{\lambda_i, \ B_i + \alpha^o\}$.

According to this definition, a rate allocation is fair if sources with low bandwidth requirements are fully satisfied while sources with high requirements obtain a fair share of the capacity according to the given fairness criteria. With pooled resources, the fair allocation to a given source depends only on the rate requirement of the source and the total capacity of all resources. With owned resources, the fair rate allocation takes into consideration the capacity B_i on the primary path of source i.

As we will discuss below, rate allocations for a network with pooled and owned resources that are fair and satisfy the respective fairness criteria are uniquely defined (with respect to the values of γ_i). Further, assuming knowledge of the load of all sources and the bandwidth of all LSPs, the rate allocations can be effectively constructed. Later in this paper, we attempt to achieve the desired rate allocations in a distributed fashion via a feedback loop, and without explicit knowledge of the traffic load of the sources.

2.1   Pooled Resources

The rate allocation distributes the aggregate capacity from all LSPs across all sources, regardless of the available capacity on the primary path of a source. Hence, the aggregate capacity on all LSPs can be thought of as a single pool of resources. We refer to α^p as the fair share of this rate allocation. The fair share α^p in a network with pooled resources is given by

$$\alpha^p = \begin{cases} \frac{\sum_{i=1}^{N} B_i - \sum_{j \in U} \lambda_j}{|O|} & \text{if } O \neq \emptyset \\ \infty & \text{otherwise}, \end{cases} \qquad (1)$$

where

$$U = \{j \mid \lambda_j < \alpha^p\} \qquad \text{and} \qquad O = \{j \mid \lambda_j \geq \alpha^p\}. \qquad (2)$$

One can think of U as the set of 'underloaded' sources that can satisfy their bandwidth demands, and of O as the set of 'overloaded' sources. Then, the fair rate allocation is obtained by subtracting the bandwidth demand from underloaded sources, and then dividing the remainder by the number of overloaded sources.

If the total demand is less than the total available capacity, i.e., $\sum_{i=1}^{N} \lambda_i \leq \sum_{i=1}^{N} B_i$, then all sources are underloaded and $\alpha^p = \infty$. In the special case where all sources are greedy, i.e. $\lambda_i = \infty$ for all $i = 1, \ldots, N$, we have $U = \emptyset$ and $\gamma_i = \alpha^p = \sum_{i=1}^{N} B_i / N$ for all i.

Efficiency of this rate allocation can be verified by inspection. A proof of the efficiency and the uniqueness of this rate allocation is given in [3], which specifies a rate allocation for a shared bus metropolitan area network.

We can construct α^p as follows. Assume, without loss of generality, that the sources are ordered according to the generated load, that is, $\lambda_i \leq \lambda_j$ for $i < j$. Select $\hat{k}$ has the largest index k ($1 \leq k \leq N$) which satisfies

$$\lambda_k \leq \frac{\sum_{l=1}^{N} B_l - \sum_{l=1}^{k} \lambda_l}{N - k} . \tag{3}$$

Then, we can determine the fair share α^p by

$$\alpha^p = \begin{cases} \frac{\sum_{l=1}^{N} B_l - \sum_{l=1}^{\hat{k}} \lambda_l}{N - \hat{k}} & \text{if } \hat{k} < N \\ \infty & \text{otherwise} . \end{cases} \tag{4}$$

2.2 Owned Resources

Here, each source may consume all of the capacity on its primary path and, in addition, a fair share of the remaining unused capacity on all secondary paths. Hence, since each source can always consume all the resources on its primary path, the capacity of the primary path can be thought of as being 'owned' by the source. For the fairness definition, we distinguish between flows that use the entire bandwidth on the primary path and those that do not:

$$\tilde{U} = \{j \mid \lambda_j < B_j\} \quad \text{and} \quad \tilde{O} = \{j \mid \lambda_j \geq B_j\}. \tag{5}$$

Thus, the total surplus capacity, which can be distributed to sources in $\tilde{O}$, amounts to $C' = \sum_{i \in \tilde{U}} (B_i - \lambda_i)$. For a source where the demand is not satisfied by the primary path, i.e., $i \in \tilde{O}$, we define $\lambda_i' = \lambda_i - B_i$. Now the fair share of the surplus is given by

$$\alpha^o = \begin{cases} \frac{C' - \sum_{j \in U'} \lambda_j'}{|O'|} & \text{if } O' \neq \emptyset \\ \infty & \text{otherwise,} \end{cases} \tag{6}$$

where

$$U' = \{j \in \tilde{O} \mid \lambda_j' < \alpha^o\} \quad \text{and} \quad O' = \{j \in \tilde{O} \mid \lambda_j' \geq \alpha^o\}. \tag{7}$$

The rate allocation is obtained via

$$\gamma_i = \begin{cases} \lambda_i & i \in \tilde{U} \text{ or } i \in U' \\ B_i + \alpha^o & i \in O' . \end{cases} \tag{8}$$

Thus, a source either obtains enough bandwidth to satisfy its demand, or it obtains the resources on its primary path and a fair share of the surplus. If all sources are greedy, with owned resources, we have $\tilde{U} = \emptyset$, $C' = 0$, $U' = \emptyset$, $\alpha^o = 0$, and therefore $\gamma_i = B_i$ for $i = 1, \ldots, N$.

As with pooled resources, the proofs in [3] can be used to establish the efficiency of the rate allocation. A value for α^o can be constructed as follows. Assume, without loss of generality, that the sources in $\tilde{O}$ have index $1, 2, \ldots, |\tilde{O}|$, and are ordered according to the generated load, that is, $\lambda'_i \leq \lambda'_j$ for $i < j$ and $i, j \in \tilde{O}$. Select $\hat{k}$ as the largest index k $(1 \leq k \leq |\tilde{O}|)$ which satisfies

$$\lambda'_k \leq \frac{C' - \sum_{l=1}^{k} \lambda'_l}{|\tilde{O}| - k} . \tag{9}$$

Then, we have

$$\alpha^o = \begin{cases} \frac{C' - \sum_{j=1}^{\hat{k}} \lambda'_j}{|\tilde{O}| - \hat{k}} & \text{if } \hat{k} < |\tilde{O}| \\ \infty & \text{otherwise} . \end{cases} \tag{10}$$

3   The Primary Path First (PPF) Property

For each source, the fairness and efficiency criteria presented in the previous section make statements about the total rate allocation to a source, but ignore how traffic is split between the primary path and the secondary paths. From a traffic engineering perspective, a rate allocation that transmits more traffic on the primary paths is more attractive, since routing traffic on secondary paths increases the fraction of out-of-sequence delivered packets, leading to higher administrative complexity.

To prevent the multipath routing scheme from spreading traffic across *all* available paths [21], we formulate an objective for our traffic engineering problem, which we call *Primary Path First (PPF)*. PPF refers broadly to the desire to limit the consumption of secondary paths in the MPLS network. In this paper, we focus on an instantiation of PPF where we seek to minimize the total volume of flow assigned to secondary paths.

To make the notion of PPF precise, we refer to a source and its primary path as a source-path pair. An $N \times N$ matrix M will be called a routing matrix if it describes the global assignment of path capacity to sources, i.e. $M(i, j)$ is the amount of traffic sent by source i to path j. Thus, the throughput for source i is $\gamma_i = \sum_{j=1}^{N} M(i, j)$, and the secondary traffic associated with source i is equal to $\sum_{j \neq i} M(i, j)$. The definition of an assignment of traffic that minimizes the total volume of flow assigned to secondary paths is as follows.

Definition 2. *Given a saturating rate allocation* $\{(\lambda_i, \gamma_i) : i = 1, \ldots, N\}$, *a routing matrix* M *is said to be PPF-optimal if it solves the following linear program.*

$$\min \quad \sum_{i=1}^{N} \sum_{j \neq i} M(i, j), \tag{11}$$

$$subject\ to \quad \sum_{j=1}^{N} M(i,j) = \gamma_i, \qquad \forall\, i = 1, \ldots, N,$$

$$\sum_{i=1}^{N} M(i,j) = B_j, \qquad \forall\, j = 1, \ldots, N,$$

$$M(i,j) \geq 0, \qquad \forall\, i, j = 1, \ldots, N.$$

From this definition, a routing matrix M is PPF-optimal if it achieves the given saturating allocation with minimum total volume of traffic sent along secondary paths. There is unnecessary use of secondary paths if either of the following cases hold.

- **Case 1:** There is a sequence $(i_1, i_2, \cdots, i_k)$ with $k > 2$ such that $M(i_1, i_2) > 0$, $M(i_2, i_3) > 0, \ldots, M(i_{k-1}, i_k) > 0$. We call such a sequence a *chain*.
- **Case 2:** There is a *cycle* $(i_1, i_2, \cdots, i_k)$ with $i_k = i_1$ and $k > 2$, such that $M(i_1, i_2) > 0, M(i_2, i_3) > 0, \ldots, M(i_{k-1}, i_k) > 0$.

With these cases in mind, we can devise a procedure that reduces the total amount of traffic sent on secondary paths without altering the total rate allocation γ_i of any source i. Suppose the rate allocation is saturating and we identify a chain which satisfies the condition in Case 1. We can eliminate the chain, or at least cut the chain into two smaller chains, by setting

$$M(i_s, i_{s+1}) \leftarrow M(i_s, i_{s+1}) - \min\{M(i_t, i_{t+1}) : t = 1, \ldots, k-1\}, \quad s = 1, \ldots, k-1,$$

$$M(i_s, i_s) \leftarrow M(i_s, i_s) + \min\{M(i_t, i_{t+1}) : t = 1, \ldots, k-1\}, \quad s = 2, \ldots, k-1,$$

$$M(i_1, i_k) \leftarrow M(i_1, i_k) + \min\{M(i_t, i_{t+1}) : t = 1, \ldots, k-1\}.$$

Suppose we identify a cycle which satisfies the condition in Case 2. Then, we can eliminate the cycle by setting

$$M(i_s, i_{s+1}) \leftarrow M(i_s, i_{s+1}) - \min\{M(i_t, i_{t+1}) : t = 1, \ldots, k-1\},$$

$$M(i_s, i_s) \leftarrow M(i_s, i_s) + \min\{M(i_t, i_{t+1}) : t = 1, \ldots, k-1\},$$

for all $s = 1, \ldots, k-1$. By adjusting the routing matrix M in this fashion, the cycle disappears, and the total volume of secondary-path traffic is reduced. However, some new shorter chain might be created.

By repeating the above steps of eliminating cycles and chains, we reduce the total volume of secondary traffic, and, eventually, obtain a routing matrix where no chains or cycles exist that satisfy the conditions of Case 1 or 2.

The following proposition shows that a necessary and sufficient condition for a routing matrix to be PPF-optimal is the absence of chains or cycles that satisfy the conditions of Cases 1 or 2.

Proposition 1. *Given a saturating rate allocation* $\{(\lambda_i, \gamma_i) : i = 1, \ldots, n\}$. *A routing matrix* M^* *that achieves this allocation is PPF-optimal if and only if there does not exist a chain or cycle as defined above.*

Proof: (Sufficiency) Consider any routing matrix M that achieves the saturating rate allocation $\{(\lambda_i, \gamma_i) : i = 1, \ldots, N\}$. Since $M(i, i) \leq B_i$ and $\sum_{j=1}^{N} M(i, j) = \gamma_i$, we have that $\sum_{j \neq i} M(i, j) \geq \max\{\gamma_i - B_i, 0\}$. Summing this lower bound across all sources, we have that

$$\sum_{i=1}^{N} \sum_{j \neq i} M(i, j) \geq \sum_{i=1}^{N} \max\{\gamma_i - B_i, 0\}. \tag{12}$$

Now consider the routing matrix M^* described in the statement of the proposition. Since there exist no chains or cycles of source-path pairs that satisfy Case 1 or 2, we can follow that, if a source i sends secondary flow, then path i does not receive secondary flow from any other source. Thus, $\sum_{k \neq i} M^*(k, i) = 0$. Since the rate allocation is saturating, we have that $B_i = \sum_{k=1}^{N} M^*(k, i)$, which implies $M^*(i, i) = B_i$, therefore $\sum_{j \neq i} M^*(i, j) = \gamma_i - B_i = \max\{\gamma_i - B_i, 0\}$. Conversely, if path i receives secondary flow from some other source, then source i itself does not send secondary flow. Then, $\sum_{j \neq i} M^*(i, j) = 0$ and $\gamma_i = M^*(i, i) \leq B_i$. Thus, $\sum_{j \neq i} M^*(i, j) = 0 = \max\{\gamma_i - B_i, 0\}$. Again, since the rate allocation is saturating, at least one of the two cases above holds for each $i = 1, \ldots, N$. Thus,

$$\sum_{i=1}^{N} \max\{\gamma_i - B_i, 0\} = \sum_{i=1}^{N} \sum_{j \neq i} M^*(i, j), \tag{13}$$

which, combined with Equation (12), implies that

$$\sum_{i=1}^{N} \sum_{j \neq i} M(i, j) \geq \sum_{i=1}^{N} \sum_{j \neq i} M^*(i, j). \tag{14}$$

Consequently, M^* is PPF-optimal with respect to the saturating rate allocation $\{(\lambda_i, \gamma_i) : i = 1, \ldots, N\}$.

(Necessity) Suppose M is PPF-optimal. If there exists a chain or cycle of source-path pairs which satisfies the conditions of either Case 1 or Case 2, then we can reduce the total volume of secondary traffic by reducing the length of the chain or by eliminating the cycle. However, this would contradict the fact that M is PPF-optimal. Thus, PPF-optimality implies no chains or cycles of source-path pairs satisfying the conditions of Case 1 or 2. ∎

From the proof, if there is a routing matrix M that achieves the saturating rate allocation $\{(\lambda_i, \gamma_i) : i = 1, \ldots, N\}$, then a lower-bound of the total secondary-path traffic is $\sum_{i=1}^{N} \max\{\gamma_i - B_i, 0\}$. From the definition of the PPF criterion, we obtain the following corollary:

Corollary 1. *A routing matrix M is PPF-optimal if it satisfies*

$$\sum_{i=1}^{N} \sum_{j \neq i} M(i, j) = \sum_{i=1}^{N} \max\{\gamma_i - B_i, 0\}. \tag{15}$$

4   Multipath-AIMD

Additive-Increase Multiplicative-Decrease (AIMD) feedback algorithms are used extensively for flow and congestion control in computer networks [5,6,11,10,12,19] and are widely held to be both efficient and fair in allocating traffic to network paths. These algorithms adjust the transmission rate of a sender based on feedback from the network following an *additive increase/multiplicative decrease* rule. If the network is free of congestion, the transmission rate of the sender is increased by a constant amount. If the network is congested, the transmission rate is reduced by an amount that is proportional to the current transmission rate. Note that in earlier instantiations of the AIMD rule the sending rate for a given source is adjusted as though only one path exists for end-to-end communication.

In this section, we generalize the AIMD rule to account for multiple paths between the sender and the receiver. The resulting algorithm, called *multipath-AIMD*, is intended to provide an efficient and fair mechanism for allocating bandwidth in an MPLS network. We assume that each LSP in the network periodically sends binary congestion-state information to all sources, similar to the DECbit scheme [12]. In the following, we develop two versions of the multipath-AIMD algorithm: basic multipath-AIMD (cf. Section 4.1) and multipath-AIMD with PPF correction (cf. Section 4.2). In the basic multipath-AIMD algorithm, each source uses the original AIMD rule to periodically adjust the rate at which it sends traffic along each LSP, providing a simple, distributed scheme for allocating network paths. Our simulation experiments indicate that the basic scheme is very robust, generally converging to an efficient and fair rate allocation within a reasonable interval of time. One undesirable feature associated with basic multipath-AIMD, especially in the case of greedy sources, is that it tends to allocate flow from all sources to all paths, completely ignoring the PPF criterion. Multipath-AIMD with PPF correction seeks to address this issue by modifying the AIMD adjustment on each LSP according to additional binary feedback information which informs sources of opportunities for reducing secondary path traffic.

4.1   Basic Multipath-AIMD

The basic multipath-AIMD algorithm consists of two parts: a feedback mechanism provided by the network and a rate adjustment mechanism implemented by the sources. The feedback mechanism is similar to the DECbit scheme [12]. Each LSP $j = 1, \ldots, N$ periodically sends messages to all sources containing a binary signal $f_j = \{0, 1\}$ indicating its congestion state. Congestion is defined in terms of the capacity B_j of LSP j: if the utilization of path j meets or exceeds B_j, then the source will receive a signal $f_j = 1$; otherwise the source receives a signal $f_j = 0$. We assume that the source receives signals on the congestion state from each path at regular intervals of length $\Delta_{LSP} > 0$ (a parameter), asynchronously with respect to all other paths.

The rate adjustment mechanism is based upon the original AIMD algorithm [13] with a slight modification to account for non-greedy sources. Each source updates its sending rates to all paths at regular intervals of length $\Delta_{src} > 0$ (also a parameter), asynchronously with respect to all other sources. In this mechanism, the most recent feedback signals received and stored by a source are used in the rate adjustments. We usually set $\Delta_{src} = \Delta_{LSP}$ so that a feedback signal is used by a source only one time. Let x_{ij} denote the rate at which source i sends traffic to path j. The formula for the adjustment depends upon whether resources are pooled or owned, as follows.

Pooled Resources. Each source i adjusts its sending rate to LSP j based on the received feedback signals according to

$$
x_{ij} \leftarrow \begin{cases} x_{ij} + k_a, & \text{if } \sum_{l=1}^{N} x_{il} < \lambda_i \text{ and } f_j = 0, \\ x_{ij} & \text{if } \sum_{l=1}^{N} x_{il} \geq \lambda_i \text{ and } f_j = 0, \\ x_{ij} \cdot (1 - k_r) & \text{if } f_j = 1. \end{cases} \tag{16}
$$

where $k_a > 0$ and $k_r \in (0, 1]$ are the additive increase and multiplicative decrease parameters, respectively, and where f_j is the latest congestion signal received for LSP j.

Owned Resources. There are two cases to consider. First, if the desired sending rate of source i does not exceed the capacity of its primary path, i.e. $\lambda_i \leq B_i$, then it adjusts its rate to LSP i according to

$$
x_{ii} \leftarrow \begin{cases} \min\{x_{ii} + k_a, \lambda_i\}, & \text{if } x_{ii} \leq \lambda_i, \\ x_{ii} \cdot (1 - k_r), & \text{if } x_{ii} > \lambda_i. \end{cases} \tag{17}
$$

Note that no flow is ever assigned from source i to any other LSP $j \neq i$. For sources i that demand more than the capacity of their primary paths, i.e. $\lambda_i > B_i$, then, after receiving feedback signals from all paths, source i adjusts its sending rate according to

$$
x_{ii} \leftarrow \min\{x_{ii} + k_a, B_i\}, \tag{18}
$$

and for $j \neq i$

$$
x_{ij} \leftarrow \begin{cases} x_{ij} & \text{if } (x_{ii} < B_i) \text{ or } (x_{ii} = B_i , \sum_{l=1}^{N} x_{il} \geq \lambda_i \text{ and } f_j = 0), \\ x_{ij} + k_a, & \text{if } x_{ii} = B_i , \sum_{l=1}^{N} x_{il} < \lambda_i \text{ and } f_j = 0, \\ x_{ij} \cdot (1 - k_r) & \text{if } x_{ii} = B_i , \text{ and } f_j = 1. \end{cases} \tag{19}
$$

Thus, in the owned case, a source never sends traffic to secondary paths before it makes full use of its primary path. Moreover, the traffic sent from a source to its primary path is independent of traffic sent from other sources.

4.2 Multipath-AIMD with PPF Correction

In the case of owned paths, the basic multipath-AIMD algorithm requires all sources to consume first the capacity of their respective primary paths, and secondary paths are utilized only when the primary paths are insufficient to meet the desired sending rate. As a result, the basic multipath-AIMD algorithm automatically produces PPF-optimal routing matrices. However, this is not the case for pooled resource. Therefore, to enforce the PPF criterion for pooled resources, we develop an alternative algorithm, called multipath-AIMD with PPF correction.

As with the basic scheme, multipath-AIMD with PPF correction consists of a feedback mechanism and a rate adjustment mechanism. As before, multipath-AIMD with PPF correction takes binary feedback $f_j = \{0, 1\}$ from all LSPs. However, in this case, extra feedback information is required to allow the sources to coordinate in attempting to reduce the total volume of secondary traffic. Each source $i = 1, \ldots, N$ periodically sends messages to all other sources containing a binary (routing) vector $m_i = (m_{i1}, \ldots, m_{iN})$, where $m_{ij} = 1$ if source i is currently sending traffic to LSP j, and $m_{ij} = 0$ otherwise. Each source i retains the routing vector m_j associated with all other sources and uses this information to modify the basic AIMD update of Equation (16). Each source transmits its routing vector at regular intervals of length $\Delta_{PPF} > 0$ (a parameter), asynchronously with respect to all other sources.

The rate adjustment mechanism for multipath-AIMD with PPF correction includes all of the rate adjustments from the basic scheme (i.e. updates of the type in Equation (16)) plus extra adjustments based on routing vector information received from the other sources. In particular, after each basic multipath-AIMD rate adjustment, each source i will engage in a PPF correction step as follows.

$$x_{ij} \leftarrow \begin{cases} \max\{x_{ij} - K, 0\} & \text{if } \sum_{l \neq i} m_{li} > 0, \\ x_{ij} & \text{otherwise,} \end{cases} \tag{20}$$

$$x_{ii} \leftarrow x_{ii} + \sum_{j \neq i} \min\{K, x_{ij}\}, \tag{21}$$

where $K > 0$ is the additive PPF correction parameter. Thus, if source i is making use of the secondary LSP $j \neq i$ and if LSP i is receiving secondary flow from some other source, then source i will reduce traffic on the secondary LSP j by K and, at the same time, increase its traffic to the primary LSP i by K.

Equations (20) and (21) have the effect of reducing flow along chains or cycles of source-path pairs with unnecessary secondary path flow. In fact, the PPF correction is inspired from the secondary path reduction scheme discussed in Section 3, which involves breaking chains or cycles of source-path pairs that satisfy either Case 1 or 2 from Section 3. By reducing the total flow along each chain or cycle with unnecessary secondary path utilization, the PPF correction creates the opportunity for subsequent basic multipath-AIMD rate adjustments to modify the solution toward efficiency, fairness, and PPF-optimality. Unfortunately, while the PPF correction creates this opportunity, it does not represent a complete solution. The basic problem is that the PPF correction tends to push flow onto primary paths, interfering with the natural tendency of AIMD to arrive at a fair distribution of the load. In fact, the basic multipath-AIMD rate adjustment (see

Equation (16)) and the PPF correction (see Equations (20)-(21)) tend to compete with one another as the system evolves to a final rate allocation. In practice, one must be careful in choosing values for k_a, k_r, and K. If K is large compared to k_a and k_r, then the PPF correction will dominate, and the resulting rate allocation will show low utilization on the secondary paths, but may also be quite far from the fair-share rate allocation. The simulation results in Section 5 illustrate this tradeoff.

5   Simulation Results

Here we present *ns-2* [1] simulation results to evaluate multipath-AIMD as applied to an MPLS network with five sources and five LSPs. In Section 5.1 we present results for the basic multipath-AIMD algorithm, and in Section 5.2 we present results for the revised algorithm, multipath-AIMD with PPF correction. Our simulations indicate that (1) the basic algorithm achieves an efficient and fair allocation of capacities to sources, (2) the basic algorithm yields a PPF-optimal solution in the case of owned resources, and (3) the revised algorithm, multipath-AIMD with PPF correction, can reduce secondary path utilization in the pooled resources case at the expense of reduced fairness.

Experimental Setup. Figure 3 illustrates the topology of the MPLS network simulated in *ns-2*. The nodes S1, S2, S3, S4, and S5 are source nodes and I1, I2, I3, I4 and I5 are ingress nodes of an MPLS network. The LSPs associated with the ingress nodes have bandwidths $(B_i \mid i = 1, \dots, 5) = (50, 40, 30, 30, 30)$ Mbps. We modified the ns-2 code in two ways. First, ingress nodes periodically send feedback messages to the sources indicating the congestion state of the corresponding LSPs, as described in Section 4. Second, source nodes generate CBR traffic with a rate specified by our multipath-AIMD scheme in response to the received feedback messages. The bandwidth and propagation delay for each link between the source nodes and the ingress nodes are set to 100 Mbps and 5 ms, respectively. Resources send congestion feedback every $\Delta_{LSP} = 5$ ms. Sources update their sending-rates every $\Delta_{src} = 5$ ms. For the experiments involving multipath-AIMD with PPF correction, the topology of Figure 3 is augmented to include full-duplex links between all source pairs, with bandwidth and propagation delay of 100 Mbps and 1 ms, respectively. Sources exchange binary routing vectors every $\Delta_{PPF} = 5$ ms. All packets in the simulation are 50 bytes in length and are treated as UDP packets (i.e. no flow control). Finally, for all experiments in this section, we set $k_a = .1$ Mbps and $k_r = .01$ as values for the additive increase and multiplicative decrease parameters, respectively.

5.1   Basic Multipath-AIMD

Experiment 1: Basic Multipath-AIMD with Greedy Sources and Pooled Resources. Figure 4 shows the outcome from basic multipath-AIMD applied to the case of greedy sources and pooled resources. The plot shows the evolution of the throughput γ_i achieved

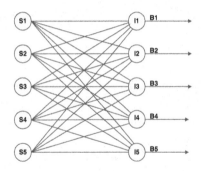

Fig. 3. Simulated network topology.

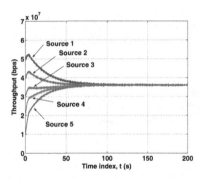

Fig. 4. Experiment 1 – Basic multipath-AIMD applied with greedy sources and pooled resources.

Table 1. Experiment 2 – Predicted efficient and fair rate allocations for the pooled and owned cases. The predictions are based on the results from Sections 2 and 3. All values are expressed in Mbps. Note that the desired sending rate for Source 2 changes from 30 to 50 at time $t = 80$ sec.

	Initial Scenario, $t \in [0, 80)$ sec			Final Scenario, $t \in [80, 200)$ sec			
Source	Load	Throughput, γ_i	Throughput, γ_i	Load	Throughput, γ_i	Throughput, γ_i	Capacity
i	λ_i	(Pooled)	(Owned)	λ_i	(Pooled)	(Owned)	of LSP i
1	10	10	10	10	10	10	50
2	30	30	30	50	46.7	50	40
3	50	50	50	50	46.7	45	30
4	60	60	60	60	46.7	45	30
5	30	30	30	30	30	30	30

by each source i. Note that all sources converge within 90 seconds to a throughput of 36 Mpbs, the fair-share allocation for this case. We point out that the final routing matrix achieved by basic multipath-AIMD is not PPF-optimal. The total volume of secondary traffic in the final resource allocation (which is not shown in a graph) is 143.5 Mbps. This is much larger than the fair PPF-optimal allocation which requires only 18 Mbps of traffic on secondary paths.

Experiment 2: Basic Multipath-AIMD with Non-Greedy Sources. Here we consider the case of non-greedy sources, and we apply the basic algorithm, where the capacities of the paths are either pooled or owned. In this experiment, the desired sending rates $(\lambda_i \mid i = 1, \ldots, 5)$ for the sources all start out at values (10, 30, 50, 60, 30) Mbps. At time $t = 80$ sec, source 2 switches its desired sending rate from $\lambda_2 = 30$ Mbps to 50 Mbps. The theoretical fair-shares for all sources (before and after the switch) are shown in Table 1, with results for both the pooled and user-owned cases. Figures 5 and 6

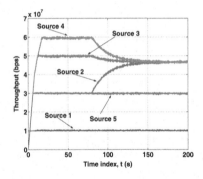
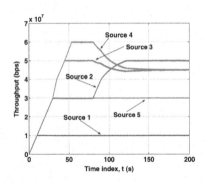

Fig. 5. Experiment 2 – Basic Multipath-AIMD applied to the case of pooled resources.

Fig. 6. Experiment 2 – Basic Multipath-AIMD applied to the case of owned resources.

illustrate the evolution of the algorithm in terms of throughput achieved by each source. The results in Figure 5 apply to the case of pooled resources, whereas Figure 6 describes the outcome for owned resources. In both figures the throughput values for each source converge to the appropriate theoretic fair-share value listed in Table 1, both before and after the switch in λ_2 at $t = 80$ sec. As before, the figures do not indicate the total volume of secondary traffic associated with the final routing matrices at $t = 200$ sec. It turns out that the final solution for the pooled resources is not PPF-optimal, with a total volume of secondary path traffic equals to 146.4 Mbps which is larger than the PPD-optimal value of 40 Mbps for the corresponding fair rate allocation for pooled resources. On the other hand, the final solution for the owned case is PPF-optimal, achieving the optimal secondary path utilization value of 40 Mbps for the fair allocation to owned resources.

5.2   Multipath-AIMD with PPF Correction

In this subsection, we present results from two experiments where we apply the revised algorithm, multipath-AIMD with PPF correction, to an MPLS network with non-greedy sources and pooled resources. In both experiments, the desired sending rates $(\lambda_i \mid i = 1, \ldots, 5)$ are (10, 50, 50, 60, 30) Mbps, and the corresponding fair-share rate allocation appears under the "Final Scenario" heading in Table 1. In Experiment 3, we set the PPF correction parameter K to a very small numerical value, $K = .00001$ Mbps, which, like basic multipath-AIMD, results in a fair but PPF-suboptimal routing matrix. In Experiment 4, we set the PPF correction parameter more aggressively, $K = .01$ Mbps, resulting in a final routing matrix which is PPF-optimal. Here, however, the corresponding rate allocation deviates from the fair-share rate allocation predicted in Table 1.

Experiment 3: Multipath-AIMD with PPF correction with $K = .00001$ Mbps. The results of this experiment are shown in Figures 7 and 8. Figure 7 shows the evolution

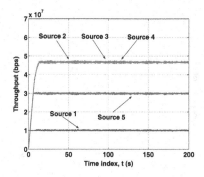

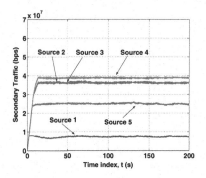

Fig. 7. Experiment 3 – Throughput achieved by multipath-AIMD with PPF correction ($K = .00001$ Mbps).

Fig. 8. Experiment 3 – Secondary traffic achieved by multipath-AIMD with PPF correction ($K = .00001$ Mbps).

of throughput for each of the five sources. The network settles within 25 seconds to a rate allocation consistent with the predicted values in Table 1. Figure 8 shows the total allocation of each source to secondary paths. Note that the final routing matrix results in 145.7 Mbps total secondary traffic, which is larger than the PPF-optimal value of 40 Mbps. Evidently, because K is so small, the PPF correction in this experiment does not have much impact in guiding the system to a PPF-optimal routing matrix.

Experiment 4: Multipath-AIMD with PPF correction with $K = .01$ Mbps. Here we apply the revised algorithm, multipath-AIMD with PPF correction, to the same problem as in Experiment 3, with a more aggressive value for the PPF correction parameter, $K = .01$ Mbps. The results are shown in Figures 9 and 10. The evolution of throughput γ_i for each of the five sources appears in Figure 9. We observe that the final rate allocation deviates from the predicted allocation from Table 1, especially with regard to sources 2, 3, and 4. Thus, the PPF correction causes the system to evolve to an unfair distribution of path resources. On the other hand, the final routing matrix is PPF-optimal with respect to the final rate allocation. This can be seen in Figure 10 which shows the evolution of secondary traffic allocated by each of the five sources. Note that the final routing matrix involves a 40 Mbps total secondary traffic, which is PPF-optimal for the rate allocation given in Figure 9.

6   Conclusions and Future Work

We have studied the problem of allocating LSP resources in an MPLS network. Our network model assumes that each traffic source has a primary path and may utilize the capacity of other, secondary, paths. We accommodate both greedy and non-greedy traffic sources and allow the capacity of each LSP to be considered as either a shared ("pooled") resource or as a resource "owned" by the corresponding traffic source.

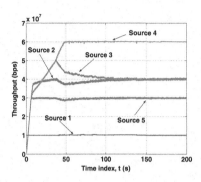
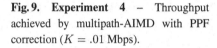

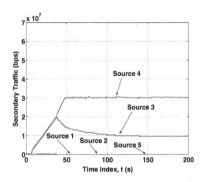

Fig. 9. Experiment 4 – Throughput achieved by multipath-AIMD with PPF correction ($K = .01$ Mbps).

Fig. 10. Experiment 4 – Secondary traffic achieved by multipath-AIMD with PPF correction ($K = .01$ Mbps).

With regard to the allocation of resources, we have defined fairness criteria based on the notion of fair-share allocation in [3], with special consideration as to whether the resources are pooled or owned. In addition to the fairness criteria, we have also introduced a secondary objective, the PPF criterion, to be achieved in the final allocation of resources. The PPF criterion is defined with respect to a given throughput allocation as an optimization model where the objective is to minimize the total volume of traffic sent along secondary paths. We provide a characterization of the PPF solution in terms of the existence of secondary path chains and cycles, and in principle this provides an algorithm that can minimize the total volume of secondary path traffic without affecting the throughput of each source.

To achieve a fair and PPF-optimal rate allocation in a distributed fashion, we propose multipath-AIMD as an extension to the earlier work of [13]. Multipath-AIMD comes in two flavors: (1) basic multipath-AIMD, which seeks to provide a fair allocation of throughput to each source, without special consideration of the PPF criterion, and (2) multipath-AIMD with PPF correction, which augments the basic algorithm to reduce the volume of secondary path traffic. Both algorithms rely upon binary feedback information regarding the congestion state of each of the LSPs and, for the second version of the algorithm, a binary routing vector associated with each source. Simulation experiments with multipath-AIMD show that the basic algorithm converges to an efficient and fair allocation of resources and also yields a PPF-optimal solution in the case of owned resources. The revised algorithm, multipath-AIMD with PPF correction, can reduce secondary path utilization (for the pooled resources case) at the expense of fairness. From the perspective of Internet traffic engineering, multipath-AIMD seems to provide a practical mechanism for improving the utilization of LSP resources, while maintaining fairness and minimizing the complexity associated with multipath routing.

This paper presents a step towards a definition of traffic engineering criteria for MPLS networks. While these initial results are promising, there are limitations to our model which must be addressed in subsequent research. First, we assume that all sources have access to all LSPs, which is clearly unrealistic in many networking contexts. This is more than an assumption of convenience, since the appropriate notion of fairness for the case where each source has access to a subset of the resources is somewhat unclear. A second limitation of our network model is that it assumes that the flows along LSPs do not interact. While it is possible to rationalize the simplified model of Figure 2, future work in this area should address the full set of interactions possible in Figure 1.

References

1. *ns-2* network simulator. http://www.isi.edu/nsnam/ns/.
2. O. Aboul-Magd, L. Andersson, and P. Ashwood-Smith. Constraint-based LSP setup using LDP. http://www.ietf.org/internet-drafts/draft-ietf-mpls-cr-ldp-05.txt, February 2001.
3. I. F. Akyildiz, J. Liebeherr, and A. Tantawi. DQDB+/-: A fair and waste-free media access protocol for dual bus metropolitan networks. *IEEE Transactions on Communications*, 41(12):1805–1815, December 1993.
4. D. O. Awduche, A. Chiu, A. Elwalid, I. Widjaja, and X. Xiao. Overview and principles of Internet traffic engineering. http://www.ietf.org/internet-drafts/draft-ietf-tewg-principles-02.txt, November 2001.
5. F. Bonomi and W. Fendick. The Rate-Based Flow Control Framework for the Available Bit Rate ATM Service. *IEEE Network*, 9(2):25–39, March/April 1995.
6. D. Chiu and R. Jain. Analysis of the Increase and Decrease Algorithms for Congestion Avoidance in Computer Networks. *Computer Networks and ISDN Systems*, 17:1–14, 1989.
7. A. Elwalid, C. Jin, S. Low, and I. Widjaja. MATE: MPLS adaptive traffic engineering. In *Proceedings of IEEE INFOCOM 2001*, volume 3, pages 1300–1309, 2001.
8. D. O. Awduche et. al. RSVP-TE: Extensions to RSVP for LSP tunnels. http://www.ietf.org/internet-drafts/draft-ietf-mpls-rsvp-lsp-tunnel-09.txt, August 2001.
9. P. Hurley, J.-Y. Le Boudec, and P. Thiran. A note on the fairness of additive increase and multiplicative decrease. In *Proceedings of ITC-16*, Edinburgh, UK, June 1999.
10. V. Jacobson. Congestion avoidance and control. In *Proceedings of ACM Sigcomm '88, August, 1988*, pages 314–329, 1988.
11. R. Jain. Congestion control and traffic management in ATM networks: Recent advances and a survey. *Computer Networks and ISDN Systems*, 28(13):1723–1738, October 1996.
12. R. Jain and K. K. Ramakrishnan. Congestion avoidance in computer networks with a connectionless network layer: Concepts, goals and methodology. *Proceedings of the Computer Networking Symposium; IEEE; Washington, DC*, pages 134–143, 1988.
13. R. Jain, K. K. Ramakrishnan, and D.-M. Chiu. Congestion avoidance in computer networks with a connectionless network layer. December 1988. Digital Equipment Corporation, Technical Report DEC-TR-506.
14. F. P. Kelly. Charging and rate control for elastic traffic. *European Transactions on Telecommunications*, 8:33–37, 1997.
15. F. P. Kelly, A. K. Maulloo, and D. K. H. Tan. Rate control for communication networks: Shadow prices, proportional fairness and stability. *Journal of the Operational Research Society*, 49:237–252, 1998.

16. S. Kunniyur and R. Srikant. End-to-end congestion control: Utility functions, random losses and ECN marks. In *Proceedings of IEEE INFOCOM 2000*, pages 1323–1332, March 2000.

17. K.-W. Lee, T.-E. Kim, and V. Bharghavan. A comparison of end-to-end congestion control algorithms: the case of AIMD and AIPD. In *Proceedings of IEEE Globecom 2001*, San Antonio, Texas, November 2001.

18. L. Massoulie and J. Roberts. Bandwidth sharing: Objectives and algorithms. In *Proceedings IEEE INFOCOM 1999*, New York, March 1999.

19. K. K. Ramakrishnan and R. Jain. A Binary Feedback Scheme for Congestion Avoidance in Computer Networks. *ACM Transactions on Computer Systems*, 8(2):158–181, 1990.

20. E. Rosen, A. Viswanathan, and R. Callon. Multiprotocol label switching architecture. draft-ietf-mpls-arch-07.txt, ftp://ftp.isi.edu/in-notes/rfc3031.txt, January 2001.

21. M. Vojnovic, J.-Y. Le Boudec, and C. Boutremans. Global fairness of additive-increase and multiplicative-decrease with heterogeneous round-trip times. In *Proceedings of IEEE INFOCOM 2000*, volume 3, pages 1303–1312, 2000.

Performance Analysis
of IP Micro-mobility Handoff Protocols[1]

Chris Blondia[1], Olga Casals[2], Peter De Cleyn[1], and Gert Willems[1]

[1] University of Antwerp
Universiteitsplein 1, B-2610 Antwerp – Belgium
{chris.blondia,peter.decleyn,gert.willems}@ua.ac.be
[2] Polytechnic University of Catalunia
Gran Capitan, Mod. D6, E-08071 Barcelona – Spain
olga@ac.upc.es

Abstract. Micro-mobility protocols have been proposed to provide seamless local mobility support. This paper focuses on the performance of the handoff schemes of two candidates for micro-mobility protocols, namely HAWAII and Cellular IP. For each handoff scheme, a simple analytical model is developed for the evaluation of two characteristic performance measures: the packet loss probability during handoff and the extra delay experienced by packets that are involved in the handoff. Application of these models allows a comparison of two important handoff schemes: the Multiple Stream Forwarding scheme of HAWAII and the Semi-soft Handoff scheme of Cellular IP.

1   Introduction

Handheld computing devices, such as palmtop computers are becoming the platform of choice for nowadays personal applications. With the evolution of these devices from a limited communication support, typical point-to-point interfaces (PSTN modem or RS-232 cable), towards high-speed packet radio access interfaces, the demand for network access to mobile users will grow exponentially. The wireless network access infrastructure will have to support a variety of applications and access speeds that should result in a service with the same level of quality as wireline users. Higher speed can be achieved in a cellular network by considering smaller cells. However, the smaller the cells are, the higher the frequency of handoffs may be. A number of access networks use network-dependent solutions for mobility management (e.g. GPRS). In this paper, we focus on Layer 3 solutions only.

Mobile IP [1], the current support of mobility in IP networks, delivers packets to a temporary address assigned to the mobile host at its current point of attachment. This temporary address is communicated to a possibly distant Home Agent. This approach applied to an environment with frequent handoffs may lead to high associated signalling load and unacceptable disturbance to ongoing sessions in terms of handoff latency and packets losses. Therefore, a hierarchical mobility management approach has been proposed (see e.g. [2]), where Mobile IP supports wide area mobility (e.g.

---

[1] Part of this work was supported by the EU, under project IST 11591, MOEBIUS.

G. Carle and M. Zitterbart (Eds.): PfHSN 2002, LNCS 2334, pp. 211–226, 2002.

mobility between different operators) while local mobility is handled by more optimized micro-mobility protocols. These protocols should incorporate a number of important design features related to location management, routing and handoff schemes. They should fulfil requirements such as simplicity to implement, scalability with respect to the induced signalling, efficiency and performance with respect to packet loss and introduced delay. This paper focuses on only one aspect, namely performance evaluation of handoff schemes.

More specifically we evaluate the performance of two prominent solutions for micro-mobility support, namely HAWAII ([3],[4]) and Cellular IP ([5],[6],[7]). For each protocol we develop an analytical model that allows computing characteristic performance measures of the handoff scheme that is used. These measures are related to packet loss and experienced delay. The models that are proposed are simple M/M/1 queueing networks that incorporate propagation delays between routers and processing times within routers. The models are not developed for dimensioning purposes, but mainly to investigate the influence of important design parameters and to compare the solutions. For this reason we have assumed Poisson background traffic and exponential processing times. The accuracy of this approximation has been discussed in [8] for the HAWAII protocol. The analytical model has been validated through simulation showing the accuracy of the model. Therefore, a validation is omitted in this paper. The simplicity of the model also allows the study of more general network topologies than the one considered in this paper. Remark that other issues related to these protocols are important (e.g. security) but are out of the scope of this paper.

The remainder of this paper is structured as follows. In Section 2, we present the reference network used to evaluate the handoff schemes. In Section 3, we describe the HAWAII protocol, we present the analytical model and obtain some performance measures. In Section 4 we do the same with Cellular IP. In Section 5 we compare some performance aspects of the MSF scheme of HAWAII and the Semi-soft Handoff scheme of Cellular IP. Finally Section 6 concludes the paper.

2   Reference Network

For the description of the two handoff protocols and their performance evaluation, we will use the following reference network (see Fig. 1). A Mobile Host (MH) moves between two Base Station (Old Base Station BSO and New Base Station BSN). Packets originating from the Corresponding Node (CN) reach BSO (resp. BSN) via the crossover router R0 and the intermediate router R1 (resp. R2). We introduce the following notations. Let Rx(X) denote the random variable defined as the time needed for a packet to be processed by router Rx, leaving via interface X. In other words, it denotes the time between the arrival of a packets at the router and its departure through output interface X. (Same notation applies to "routers" BSO and BSN). Furthermore, let (Rx,Ry) denote the propagation time on the link between router Rx and router Ry.

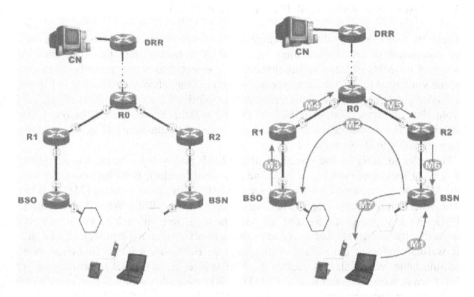

Fig. 1. Reference Network   **Fig. 2.** Messages in HAWAII MSF

3   HAWAII

In this Section we describe the handoff mechanism used in HAWAII and present an analytical model to evaluate the packet loss and delay experienced by packets involved in a handoff.

3.1   Network Model and Routing

HAWAII is a domain-based approach for supporting mobility ([3] and [4]). The gateway into each domain is called the Domain Root Router (DRR). The Mobile Host (MH) keeps it network address unchanged while moving within a domain. The Corresponding Nodes (CN) and the Home Agent (HA) do not need to be aware of the host's position within the domain. To reach the MH, HAWAII uses specialized path setup schemes that update forwarding entries in specific routers. When a router receives a packet for an unknown MH, it uses a preconfigured default interface pointing towards the domain root router. The packet will be forwarded in that direction till it arrives at a router that knows a route to the MH.

3.2   Description of the Handoff Protocol in HAWAII

There are two kinds of path setup schemes. One for networks with MHs that can only be connected to one base station (e.g. TDMA networks) and the other kind of schemes for MHs that can be simultaneously connected to two or more base stations at the same time (e.g. CDMA networks). From the four schemes considered in [3], we consider only the Multiple Stream Forwarding (MSF). This scheme forwards packets from the Old Base Station (BSO) to the New Base Station (BSN) before being diverted at the crossover router (i.e. a router where the path from CN to BSO and the path from CN to BSN cross).

In order to describe the operation of the MSF path setup scheme, we define the following messages (see Fig. 2.). At the instant of handoff, BSO loses contact with the MH and at the same time the MH sends a MIP registration message (M1) to BSN. The latter sends a path setup update message M2 to the BSO. When M2 arrives at BSO, the BSO starts to forward all packets with destination MH via router R1, including those packets that arrive after the handoff instant and that were stored in a forwarding buffer at the BSO. For that purpose, BSO adds a forwarding entry to its routing table indicating that packets for MH should leave the BSO via interface A. BSO sends the path setup message (M3) to R1, who adds a forwarding entry to its routing table indicating that packets for the MH should leave the R1 via interface A. R1 sends the path setup message (M4) to R0, who adds a forwarding entry indicating that packets for the MH should leave the R0 via interface C. From this instant on, all packets arriving at router R0 are sent directly to BSN. The path setup message continues (M5 and M6) triggering similar actions until it reaches BSN.

3.3   Performance Evaluation of HAWAII

The use of the HAWAII MSF path setup scheme implies that packets with destination the MH will follow a route, depending on their time of arrival at certain routers. In what follows, the packets are divided in classes according to the path they follow. The timing of these classes will be given from the point of view of a packet originating from the corresponding host that arrives at R0. Let t_{ho} be the time instant of handoff (i.e. the time instant that BSO loses contact with the MH and that M1 is generated in the MH). We assume that the time needed to update the forwarding entries is about equal to the service time of any packet in any router.

Class 1 packets: These packets arrive at the BSO after the handoff took place and before the new forwarding entry is added at the BSO. They arrive at the R0 after

$$t_0 = t_{ho} - \{R0(B) + (R0, R1) + R1(B) + (R1, BSO)\} \tag{1}$$

but before

$$t_1 = t_{ho} + \{(MH, BSN) + BSN(A) + (BSN, R2) + R2(A) + (R2, R0) + R0(B)\} \tag{2}$$

These packets are stored in the forwarding buffer of the BSO until they are forwarded to the MH using the route R0-R1-BSO-[forwarding buffer]-R1-R0-R2-BSN.

Class 2 packets: These packets arrive at R1 before M3 causes the adding of the new forwarding entry at R1, and they arrive at the BSO after the new forwarding entry has been added. They reach the MH via the route R0-R1-BSO-R1-R0-R2-BSN. At time instant

$$t_1' = t_1 + (R0, R1) + R1(B) + (R1, BSO) + BSO(A) + (BSO, R1) + R1(A) \tag{3}$$

message 3 is processed and router R1 changes its forwarding entries for packets with destination the MH. Hence, the packets that arrive at R0 in the interval $[t_1, t_2]$, with

$$t_2 = t_1' - \{R0(B) + (R0, R1)\} \tag{4}$$

belong to class 2.

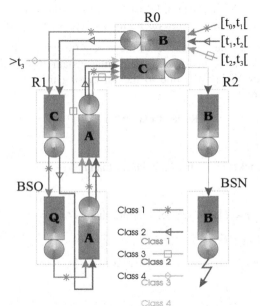

Fig. 3. A queuing model for the MSF scheme

Class 3 packets: These packets arrive at R0 before M4 causes the adding of the new forwarding entry at R0, and they arrive at R1 after the new forwarding entry has been added. They reach the MH via the route R0-R1-R0-R2-BSN. At time instant

$$t_3 = t_2 + R0(B) + (R0, R1) + (R1, R0) + R0(C) \tag{5}$$

router R0 changes its forwarding entries for packets with destination the MH (based on message M4). Therefore, packets arriving at R0 in the interval $[t_2, t_3]$, belong to class 3.

Class 4 packets: These packets are forwarded to the BSN directly, i.e. via the route R0-R2-BSN. They arrive at R0 after time instant t3.

Remark: The MSF scheme may result in the creation of routing loops (for example, after BSO has changed its entry to forward packets but before Router 1 processes M3). However these loops exist only for extremely short periods of time. In our model we only consider possible loops for packets that belong to class 1, as the probability for the occurrence of other types of loops seems to be negligible.

We derive two performance measures for this handoff scheme: the additional delay introduced by the forwarding scheme and the packet loss probability in the forwarding buffer of BSO.

Consider a stream of packets originating from the corresponding host arriving at R0 with a constant interarrival time of T ms. Let the arrival instant u of the first packet at R0 be uniformly distributed over $[t_0, T]$. In what follows we let $t_0 = 0$. Then it is possible to compute the probability distribution of the additional time introduced by the MSF scheme. In fact, this is the time needed for the packet to reach R0 (possibly after being routed via R1 and BSO). Remark that for packets arriving at R0 after time instant t_3, this additional time is zero (as no forwarding is applied anymore).

The probability that the k-th packet of the stream (arriving at R0 at time instant $(k-1) \times T + u$) arrives back in R0 later than time instant $(k-1) \times T + u + t$ (i.e. it experiences an extra delay of t due to the forwarding mechanism) is given by the following expression.

Prob[k-th packet arrives at R0 later than instant $(k-1) \times T + t + u$] =   (6)

Prob[$(k-1) \times T + u < t_1$]

  $\times$ Prob[k-th packet has extra delay larger than $t \mid (k-1) \times T + u < t_1$]

+ Prob[$t_1 \leq (k-1) \times T + u < t_2$]

  $\times$ Prob[k-th packet has extra delay larger than $t \mid t_1 \leq (k-1) \times T + u < t_2$]

+ Prob[$t_2 \leq (k-1) \times T + u < t_3$]

  $\times$ Prob[k-th packet has extra delay larger than $t \mid t_2 \leq (k-1) \times T + u < t_3$]

+ Prob[$t_3 \leq (k-1) \times T + u$]

  $\times 0$

In the above, the probability expression regarding packets of class 1 can be established as follows. It is assumed that, as soon as the new forwarding entries are set in BSO, the packets of class 1 that are stored in the forwarding buffer, are put in the queue of output interface A. Hence, they will be forwarded to R1, but when they arrive at R1 before the forwarding entries are updated, they will be sent right back to BSO. This looping occurs roughly when the waiting and service of the packet in the output queue take less time than the waiting and processing of M3 in R1. We denote the service rate of a packet μ. Then we can write:

Prob[k-th packet has extra delay larger than t | $(k - 1) \times T + u < t_1$] =
Prob[k-th packet gets in loop]
$\qquad$ (7)

$\qquad \times$ Prob[$R0(B) + (R0, R1) + R1(B) + (R1, BSO) + (t1 - (k - 1) \times T - u)$
$\qquad + (k - 1) / \mu + BSO(A) + (BSO, R1) + R1(B) + (R1, BSO) + BSO(A)$
$\qquad + (BSO, R1) + R1(A) + (R1, R0) > t \mid (k - 1) \times T + u < t_1$]
$+$ Prob[k-th packet does not get in loop]
$\qquad \times$ Prob[$R0(B) + (R0, R1) + R1(B) + (R1, BSO) + (t1 - (k - 1) \times T - u)$
$\qquad + (k - 1) / \mu + BSO(A) + (BSO, R1) + R1(A) + (R1, R0) > t \mid (k - 1)$
$\qquad \times T + u < t_1$]

Clearly, if $k = 1$ the packet has a larger probability (Prob[$BSO(A) < R1(A)$]) of getting into this routing loop than if $k = 2$ (Prob[$BSO(A) + (BSO - 1/\mu) < R1(A)$]). The looping probability of the possible other class 1 packets is considered to be negligible.

Regarding class 2 packets, we do not consider possible loops, so:

Prob[k-th packet has extra delay larger than $t \mid (k - 1) \times T + u < t_2$] =
$\qquad$ (8)

Prob[$R0(B) + (R0, R1) + R1(B) + (R1, BSO) + BSO(A) + (BSO, R1)$
$\qquad + R1(A) + (R1, R0) > t \mid t_1 \leq (k - 1) \times T + u < t_2$].

Finally, for class 3 packets we can write:

Prob[k-th packet has extra delay larger than $t \mid t_2 \leq (k - 1) \times T + u < t_3$] =
$\qquad$ (9)

Prob[$R0(B) + (R0, R1) + R1(A) + (R1, R0) > t \mid t_2 \leq (k - 1) \times T + u < t_3$].

To compute the different delay components we assume that all routers on a path that a packet follows are modelled as simple M/M/1 queueing systems. The exponential service time of a packet both includes the time needed for the routing of the packet as well as the time needed to actually put the packet on the output link. We denote the corresponding load of a router ρ. Conditioning on the value of u, the arrival instant of the first packet of the stream, all probabilities occurring in the expression for the delay packets experience due to the forwarding mechanism are of the form Prob[$X + c < t$], where X is a random variable being the sum of response times in an M/M/1 queue and c a constant. It is well known that the distribution of the response time in an M/M/1 queue is again an exponential distribution with rate $\mu(1-\rho)$. Hence the random X is a sum of exponentially distributed random variables. It is possible to find an explicit expression for the distribution of X, and hence the value of Prob[$X + c < t$] can be computed explicitly. Also some more complicated probabilities that occur in the expression, such as Prob[$X + c < t < Y + d$], Prob[$X + Y + c \leq t < X + Z + d$] or Prob[$X + c < t \mid Y < d$] ($X$, Y and Z sums of exponentially distributed variables), can be computed through some standard conditional probability techniques. Applying these general expressions for the special cases above leads to the probability that the k-th packet of the stream arrives at R0 later than instant $(k - 1) \times T + u + t$.

Now we evaluate the packet loss in the forwarding buffer in BSO. Clearly the number of packets that need to be buffered is exactly the number of packets belonging to class 1. Hence, the probability that k packets need to be stored equals

$$P((k-1)\times T + u \le t_1 < k \times T + u).$$ (10)

Using the assumption that the routers are modelled as M/M/1 queues, this probability can be computed easily.

We now want to compute the expected number of packets that will be dropped due to the expiration of a playout buffer at the MH because of the forwarding mechanism introduced by the MSF scheme. A packet will be dropped in case of a playout delay of t whenever it experiences an extra delay (due to forwarding) larger than t.

To approximate the expected number of dropped packets, we consider the different classes separately:

E[number of dropped packets] =

$$\sum_{i=1,2,3} E[\text{number of dropped packets of class } i].$$ (11)

For $i=2,3$ we have that

E[number of dropped packets of class i] = E[# class i packets] $\times p_i$,

where p_i denotes the probability that a class i packet is dropped. This probability is the same for every class i packet and is of the form $P[X + c > t]$, X being a sum of response times, so p_i can be computed.

To obtain the expected number of packets in the respective classes, we consider the following probability distributions:

Prob[# class 1 packets = k] = Prob[$(k-1) \times T + u \le t_1 < k \times T + u$],

Prob[# class 1 and 2 packets = k] = Prob[$(k-1) \times T + u \le t_2 < k \times T + u$], (12)

Prob[# class 1, 2 and 3 packets = k] = Prob [$(k-1) \times T + u \le t_3 < k \times T + u$].

These probabilities are of the same form as before and the expected value of the distributions can be easily computed. Furthermore, we have that

E[# class 2 packets] = E[# class 1 and 2 packets] - E[# class 1 packets]

E[# class 3 packets] = E[# class 1,2 and 3 packets]
                          - E[# class 1 and 2 packets]. (13)

Now to obtain an approximation for E[number of dropped packets of class 1], let n_1 denote the smallest integer larger than E[# class 1 packets]. Furthermore, let $p_{1,k}$ be the probability for the k-th packet of the stream to be dropped, given that it belongs to class 1. For every k, this probability can be computed as pointed out before. Then, with p_1 denoting the mean value of $p_{1,1}, \ldots, p_{1,n_1}$, we have (approximately) :

E[number of dropped packets of class 1] (14)

    $= n_1 \times p_1 - (n_1 - E[\text{# class 1 packet}]) \times p_{1,n_1}.$

This gives an approximation for the expected number of packets dropped in the playout buffer due to the MSF forwarding mechanism.

3.4   Numerical Results

In this section we apply the above analytical models to investigate the following performance measures. First, the packet loss probability as a function of the capacity of the forwarding buffer in BSO is determined. Next the distribution of the delay experienced by packets that arrive after handoff due to the use of the forwarding scheme is computed.

Consider the system depicted in Fig. 1, where each router (including the Base Stations) has the following characteristics: the service rate μ equals 10 packets per ms, the load is given by ρ = 0.8, and the propagation delay to each neighbouring router is variable (5 ms, 10 ms and 20 ms). We consider a stream of packets arriving at Router R0 with a constant packet interarrival time of T=20 ms.

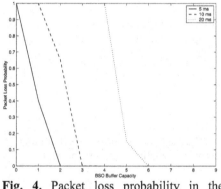

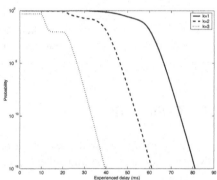

Fig. 4. Packet loss probability in the Forwarding Buffer

Fig. 5. Delay experienced by packets due to forwarding

For this system, Fig. 4. shows the packet loss probability as a function of the capacity of the forwarding buffer at the Old BS for the different propagation delays between neighbouring routers (5 ms, 10 ms and 20 ms).

Next, we compute the delay due to forwarding for this system. Fig. 5. shows the probability that the k-th packet after handoff experiences a delay of at least t ms due to the forwarding scheme.

The same evaluation is made for different values of the router service time. We assume a system where the propagation delay to the neighbouring routers is 5 ms and variable router service rate (2 packets/ms, 5 packets/ms and 10 packets/ms). Figure 6 shows the delay experienced by the first and second packet for the different values of the service rate μ.

Now we give an example where the expected number of packets that are dropped for a certain playout delay are computed, for variable router service rate μ. We consider again the system described in the previous examples, with the load of each router ρ = 0.8, the propagation delay to each neighbouring router 5 ms and variable router service rate (1 packet/ms, 2 packet/ms, 5 packet/ms and 10 packet/ms). We consider a stream of packets arriving at Router R0 with a constant packet interarrival time of T=20 ms. Fig. 7 shows the expected number of packets dropped as a function of the playout buffer for the different values of μ.

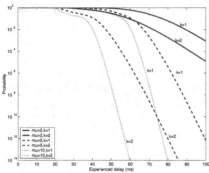 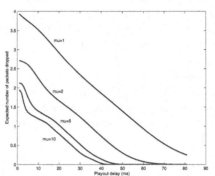

Fig. 6. Delay experienced by packets due to forwarding for variable router service rates.

Fig. 7. Expected number of packets dropped due to expiration of playout delay

4   Cellular IP

In this Section we briefly describe Cellular IP and present an analytical model to evaluate the packet loss and delay experienced by packets involved in a handoff.

4.1   Network Model and Routing

A Cellular IP ([5],[6],[7]) access network is connected to the Internet via a gateway router. A MH attached to an access network will use the IP address of the gateway as its Mobile IP care-of address. The main component of Cellular IP networks is the base station that is used as a wireless access point to which the MH relays the packets it wants to transmit and from which it receives the packets destined to it.

After power up a MH has to inform the gateway router about its current point of attachment by means of a route update message. This message is received by the base station and forwarded to the gateway on a hop-by-hop basis. Each Cellular IP node maintains a route cache in which it stores host based routing entries. The path taken by uplink packets is cached by all intermediate nodes. To route downlink packets addressed to the MH, the path used by recently transmitted packets from the MH and which have been cached is reversed.

As the routing entries in the nodes are soft state, a MH that has no data to transmit, has to send periodically special IP packets to maintain its routing path. Paging is used to route packets to MHs that have not been transmitting or receiving data for a while and whose routing cache entries have timed out. Paging caches are not maintained in each node and have a longer timeout value. If a node without a paging cache, receives a packet for a MH for which it has no routing cache entry, it will broadcast it to all its downlink neighbours.

4.2   Description of the Handoff Protocol in Cellular IP

In Cellular IP a mobile host initiates a handoff (based on signal strength measurements) by sending a route update packet to the new base station. This packet travels in a hop-by-hop manner from the base station to the gateway router and reconfigures the route cache entries in the Cellular IP nodes along its way.

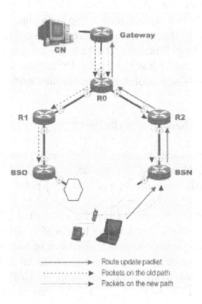

Fig. 8. Handoff in Cellular IP

Cellular IP supports three types of handoff mechanisms: *hard handoff*, *semi-soft handoff* and *indirect semi-soft handoff*.

1. *Hard Handoff:* The wireless interface of a mobile host changes from one base station to another at once. Data packets arriving at the cross-over node before the route cache entry is changed are misdirected and will be lost since the mobile host is already attached to the new base station.
2. *Semi-soft Handoff:* The mobile host switches to the new base station, transmits a route update packet with a flag indicating a semi-soft handoff while listening to the old base station. The route update packet reconfigures the route caches on the way to the gateway router and adds an entry to the cache of the crossover node. Downlink packets for the mobile host are duplicated in the crossover node and sent along both paths, the new one and the old one. After a *semi-soft delay*, the mobile host performs a regular handoff. It migrates to the new base station and sends another route update packet to complete the handoff. This last packet stops the crossover node duplicating packets. The semi-soft delay is a fixed amount of time that is proportional to the mobile-to-gateway round-trip delay. If the path to the new base station is shorter than to the old one, some packets may not reach the mobile host. To overcome this problem, packets sent along the new path need to be delayed. A delay device mechanism, located at the crossover node,

should provide sufficient delay to compensate for the time difference between the packets travelling on the old and new paths.

3. *Indirect Semi-soft Handoff:* This mechanism can be applied when using wireless technologies that have no simultaneous connection capability. They cannot listen to the old base station while sending a route update packet to the new base station. To initiate a handoff the mobile host sends the route update packet to the old base station, using as destination IP address the address of the new base station. The old base station forwards this packet with a flag indicating indirect semi-soft handoff uplink to the gateway. The gateway delivers the packet to the new base station using normal IP routing. When the new base station receives this packet, it creates a semi-soft route update packet with the IP address of the mobile host and it forwards it upstream. The rest of the algorithm then works as a semi-soft handoff.

4.3   Performance Evaluation of Cellular IP

We evaluate two performance measures for Cellular IP, namely the number of packets lost in the Hard Handoff scheme and the delay that has to be introduced in R0 to compensate for the difference in the old and new paths in the Semi-soft Handoff scheme.

First we remark that the number of packets lost in the BSO during Hard Handoff equals the number of packets that have to be stored in BSO in the MSF scheme of HAWAII. Therefore we can use the model described in Section 3.3.

In order to compute the delay that has to be introduced in R0 to compensate for the difference in old and new paths in the Semi-soft Handoff scheme, we first give an expression of the length of these paths, L_{old} and L_{new} respectively.

$$L_{old} = R0(B) + (R0, R1) + R1(B) + (R1, BSO) + BSO(B) \tag{15}$$

$$Lnew = R0(C) + (R0, R2) + R2(B) + (R1, BSN) + BSN(B)$$

Let again the arrival instant u of the first packet at R0 be uniformly distributed over $[t_0, T]$. Then the probability that the number of lost packets during handoff exceeds k is given by

$$P(\# \text{ lost packets during handoff} \geq k) \tag{16}$$
$$= P(L_{old} - u > L_{new} + (N + k - 1) \times T)$$

Since all service times in the routers are exponentially distributed, this probability can be computed easily.

4.4   Numerical Results

First we evaluate the case of the Hard Handoff scheme. Consider the reference network (Fig. 1.), where the service rate μ of each router equals 10 packets per ms, the load is given by $\rho = 0.8$, and the propagation delay to each neighbouring router is variable (5 ms, 10 ms and 20 ms). We consider a stream of packets arriving at Router R0 with a constant packet interarrival time, taking values between 5 ms and 35 ms. Fig. 9. shows the expected number of packets that are lost at the BSO due to the hard handoff, for variable packet interarrival times and different values of propagation delay. Clearly the expected number of lost packets increases when the propagation delay increases and the interarrival time decreases. Remark that the results obtained in Fig.9. correspond with those obtained in the evaluation of the number of buffers required in BSO in the HAWAII protocol.

Next, for the Semi-soft Handoff scheme we compute the packet loss probability as a function of the number of packets that are delayed in the R0 buffer before the first packet is released, in order to cope with the different propagation delay between R0-BSO and R0-BSN. We assume this difference to be 50 ms, 75 ms and 100 ms. The three curves in Fig.10 show the loss probability of packets due to early arrival in BSN for different values of the buffer capacity in R0.

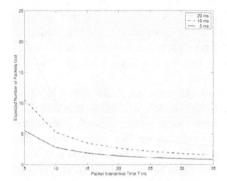

Fig. 9. Hard handoff : Expected number of packets lost in BSO

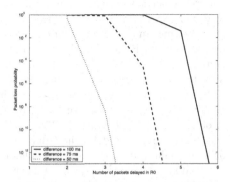

Fig. 10. Packet loss probability as a function of number of delayed packets for different values of difference in length of new and old path

5   Comparison HAWAII – Cellular IP

In this Section we use the models developed in Sections 3 and 4to compare performance aspects of the MSF scheme of HAWAII and the Semi-soft Handoff scheme of Cellular IP. In particular, we compare the number of buffers that are required in both schemes to assure a zero or negligible packet loss and we compare the delay induced by the use of these handoff schemes. Notice that this comparison is not completely fair as the mobile host's capabilities are not the same. For the Semi-soft handoff scheme in Cellular IP, the MH must be able to send a route update packet to the new BS, while listening to the old BS, whereas in the MSF scheme the MH switches instantaneously from the old BS to the new BS.

The MSF scheme of HAWAII requires buffers in order to store packets that arrive in BSO after the handoff occurs and before the forwarding entry in BSO is updated, i.e. to store temporarily class 1 packets. In Section 3 we have shown how this number of buffers can be obtained. The R0 buffers required in the Cellular IP have a different task. They are required to delay packets before sending them along the new path in order to compensate for the time difference between the old and the new path to the MH. A model that allows to evaluate this required number of buffers has been proposed in 3.3.

Consider the system presented in Fig. 1., where the routers have a service rate μ of 10 packets/ms, the load of each router is given by $\rho=0.8$, the distance between neighbouring nodes on the old route (i.e. on the route R0-R1-BSO-MH) is 20 ms and the distance between the neighbouring nodes on the new route is variable, taking values 1 ms, 3 ms, 5 ms, 10 ms and 15 ms. We consider a stream of packets arriving at Router R0 with a constant interarrival time of $T=20$ ms. Fig.11. compares the number of buffers required in BSO when using the HAWAII MSF scheme with the number of buffers required in R0 when using the Semi-soft handoff scheme of Cellular IP, in order to obtain a zero packets loss. This result shows that Cellular IP requires less buffers than HAWAII. This is due to the fact that in Cellular IP the buffers are needed to compensate for the time difference between the old path and the new path, while in HAWAII, the buffers need to accommodate packets arriving during a time equal to the sum of the old path and the new path (i.e. the time required by M2 to reach BSO). Clearly, the longer the new path, the more buffers HAWAII requires. Cellular IP on the other hand requires less buffers since a longer new path leads to less difference between the new and old path.

Next we compare the delay introduced by both schemes. Consider again the same system as defined in the above example. We use the models derived in II.B and III.B to obtain this delay. For the HAWAII MSF scheme we compute for each inter-router distance, the expected delay the k-th ($k=1,2,3,..$) packet experiences. We compute this delay for values of k for which this delay is not negligible (i.e. for all packets that are involved in the handoff). By taking the average of these values, we obtain the average delay a general packet, that is involved in the handoff, experiences. Remark that delays for different values of k may be quite different; e.g. when the distance between neighbouring routers on the new path is 10 ms, then the delay the first packet (i.e. $k=1$) experiences is 165 ms, while the delay for the eighth packet (i.e. $k=8$) is 30 ms.

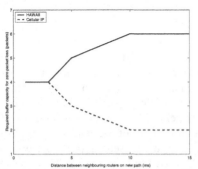

Fig. 11. Comparison of the buffers required in MSF HAWAII and Semi-soft Handoff in Cellular IP

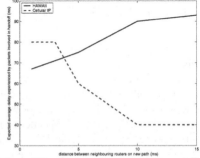

Fig. 12. Comparison of the delay experienced by packets involved in the handoff in MSF HAWAII and Semi-soft Handoff in Cellular IP

From Fig. 12., we see that in this case the average delay is given by 90 ms. For the Cellular IP scheme, the delay experienced by every packet that is involved in the handoff equals the number of buffers obtained in Fig.10. multiplied by the packet interarrival time T=20 ms. Fig. 12. compares the results for the two schemes. This figure shows that, apart from the case of very small distances between neighbouring nodes, packets in Cellular IP will experience (in average) less delay than packets in HAWAII. Remark however that in this numerical example, we have assumed that the difference in length (in ms) between the old and new path is known. In a real environment, this difference is not known and a conservative (worst case) value has to be selected, leading to possible much longer delays. In HAWAII, not all packets experience the same delay. The first packets may have a much longer delay than the last packets that are involved in the handoff. This phenomenon, which is not clear from Fig. 12., is illustrated in Fig. 5.

6   Conclusions

We have developed simple analytic models to evaluate the performance of handoff schemes of two important micro-mobility solutions, namely HAWAII and Cellular IP. The models that are proposed are simple M/M/1 queueing networks that incorporate propagation delays between routers and processing times within routers. The models are not developed for dimensioning purposes, but mainly to investigate the influence of important design parameters and to compare the two solutions. For this reason we have assumed Poisson background traffic and exponential processing times. A detailed validation of the approximations has been made in [8] by comparison with simulation. The simplicity of the model also allows the study of more general network topologies than the one considered in this paper.

The performance analysis mainly focuses on packet loss probabilities during handoff and extra packet delay introduced by the handoff schemes. For the HAWAII MSF scheme we compute the required capacity of the buffer in the old Base Station in order to have zero or negligible packet loss. Moreover we compute the delay distribution experienced by the different packets involved in the handoff. This delay can be used to compute the number of lost packets when a playout buffer is used at the destination. For the Cellular IP protocol, we derive the number of dropped packets when using a hard-handoff scheme. Moreover, for the semi-soft handoff scheme, we compute the number of packets that have to be delayed before releasing the first packet in order to compensate for the time difference between the old path and the new path.

These results are used to make a comparison of the MSF scheme in HAWAII and the semi-soft handoff scheme in Cellular IP. It is shown that, in general Cellular IP needs less buffers than HAWAII, mainly due to the fact that in Cellular IP the buffers are needed to compensate for the time difference between the old path and the new path, while in HAWAII, the buffers need to accommodate packets arriving during a time interval equal to the sum of the old path and the new path. With respect to the delay introduced during handoff, we have seen that the delay is constant for all packets involved in the semi-soft handoff scheme of Cellular IP. But as the difference in length between the old path and the new path is not known in advance, the delay that is enforced in the cross-over router may be difficult to engineer. In the MSF scheme in HAWAII on the contrary, the delay a packet experiences is variable and decreases with the sequence number of the packet involved in the handoff. This extra delay is a result of the forwarding scheme itself and does not need to be engineered.

References

1. C. Perkins, ed., "IP Mobility Support", IETF RFC 2002, October 1996.
2. R. Caceres and V. Padmanabhan, "Fast and scalable handoffs for wireless networks", in Proc. ACM MOBICOM '96, pp. 56-66, 1996
3. Ramjee, R., La Porta, T., Thuel, S., Varadhan, K., and Wang, S., HAWAII: a domain based approach for supporting mobility in wide–area wireless networks, Proceedings of International Conference on Network Protocols, ICNP'99.
4. R. Ramjee, T. La-Porta, S. Thuel, K. Varadhan, L. Salgarelli, "IP micro-mobility support using HAWAII", Internet draft, July 2000.
5. A. Valko, "Cellular IP – a new approach of Internet host mobility", ACM Computer Communication Reviews, January 1999
6. A. Campbell, J. Gomez, C. Y Wan, S. Kim, Z. Turanyi, A. Valko, "Cellular IP", IETF draft (draft-ietf-mobileip-cellularip-00.txt), January 2000.
7. A. Campbell, J. Gomez, S. Kim, A. Valko, C.-Y. Wan and Z. Turanyi, "Design, implementation and evaluation of Cellular IP", IEEE Personal Communications, August 2000, pp.42-49
8. C. Blondia, O. Casals, L. Cerda and G. Willems, "Performance analysis of a forwarding scheme for handoff in HAWAII", to appear in Proceeedings of Networking 2002, Pisa 2002, win-www.uia.ac.be/u/pats/publications.html

High-Speed Mobile and Wireless Networks

James P.G. Sterbenz*

BBN Technologies, 10 Moulton St., Cambridge, MA 02138-1191 USA,
jpgs@ieee.org

Abstract. Mobile wireless networks have significantly different characteristics and constraints from wired networks. While link rates in these networks have traditionally been lower than fixed fiber optic networks, mobile and wireless links are getting more and more sophisticated. 802.11a operates at 54Mb/s, LMDS at 36Mb/s; satellite links of 622Mb/s have been experimentally deployed, and fixed point-to-point RF and free wireless links at 100Mb/s and beyond are available.

Mobile links present challenges to high-speed networks in the control plane due to the dynamic nature of topology and the resulting impact on QOS, such as time-varying bandwidth capacity and latency. Wireless links require fundamentally different mechanisms due to their unreliability, ranging from error control mechanisms to correct bit errors to compensating for highly dynamic channel conditions. As the bandwidth-×-delay product of these links increase, so does the difficulty in dealing with these problems. Current satellite and deep space links have bandwidth-×-delay products that are no larger than terrestrial networks, but the potential for petabits or exabits exists in future free-space laser interplanetary links.

This discussion session will explore the issues and challenges posed by high speed mobile wireless networks, and ways in which to meet these challenges at the MAC, link, transport, session, and application layers.

---

* Moderator

G. Carle and M. Zitterbart (Eds.): PfHSN 2002, LNCS 2334, p. 227, 2002.
© Springer-Verlag Berlin Heidelberg 2002

Peer Networks –
High-Speed Solution or Challenge?

Joseph D. Touch*

USC/ISI

Abstract. Application, transport, and network-layer overlays are becoming a dominant force to provide ubiquitous, large-scale network architecture. These 'peer networks', such as Gnutella, Freenet, and Napster, demonstrate the power of automatic network configuration, achieving world-wide deployment of a virtual topology in a matter of weeks. This deployment simplicity comes with costs, e.g., current systems reimplement core network functions at the application layer, potentially introducing inefficiency, incompleteness, and incorrectness.

Peer to peer networks flatten and democratize the conventionally hierarchical network topology. The current Internet is based on IP addresses and Autonomous System (AS) numbers, allocated as hierarchies which rely on managed global identifiers to provide packet forwarding using local information. By contrast, peer networks often allow more local management, at the expense of forwarding efficiency. In some systems, packets take $O(\sqrt{N})$ hops (vs. $O(\log N)$ in hierarchies); in others, paths meander or even backtrack. Such expenses are tolerated in exchange for simpler, automatic configuration which enables rapid, ubiquitous deployment.

Peer networks can defeat the careful aggregation of provisioned bandwidth. The use of application or transport layer tunnels can result in traffic traversing a network-layer link multiple times. Separate links which would have been multiplexed together nearer their sources can be tunneled far into the core before being multiplexed, obscuring queuing interactions that would have been detected or reacted to more efficiently at the edge. Application layer routing can contradict network layer routing, resulting in meandering paths, or sometimes cycles.

This panel examines these aspects of peer networks, and how best to utilize their strengths to enable network scale while avoiding their potential pitfalls.

---

* Moderator

G. Carle and M. Zitterbart (Eds.): PfHSN 2002, LNCS 2334, p. 228, 2002.
© Springer-Verlag Berlin Heidelberg 2002

High Speed Networks for Carriers

Karl J. Schrodi

Siemens AG, Information and Communication Networks,
81359 Munich, Germany
karl.schrodi@icn.siemens.de

Abstract. This paper focuses on high speed networks in a future public network infrastructure. 'Next Generation Networks' (NGNs), built on fast IP based packet switching technologies, will provide a unified platform capable of seamlessly supporting a variety of existing and future telecommunications and data services and applications. Requirements on and expected properties of NGNs as the new generation of carrier networks are discussed. An architectural overview reveals the major interfaces and related protocol issues. An implementation approach with emphasis on QoS, network resilience and operational cost issues is presented.

1 Introduction

High speed networks have long played a role in campuses and other closed environments for scienctific, research and educational purposes. Support from public networks was limited to the provisioning of transmission capacities for the interconnection of different high speed networking islands. Corporate business and industrial applications relied on the same scheme, since the public network was dedicated to and optimized for telephony services, and a public network infrastructure suitable for high speed information switching was not available.

Asynchronous Transfer Mode (ATM) [1] was initially intended to become the key technology for the introduction of generally available broadband telecommunication services in local and wide area networks and "the transfer mode solution for implementing a B-ISDN" [2]. However, ATM missed the path towards mass applications and, despite of its proven capabilities of fulfilling all major telecommunication networking requirements, it may finally find itself pushed back into the role of 'just another data link layer'.

The real drive towards higher speed and higher throughput in a public networking infrastructure found its origin in a different application area, which was known and used by a rather small community of data communication insiders only even less than ten years ago. Internet and World Wide Web (WWW) [3], [4] have become an almost unbelievable success story. Internet usage and traffic volume have grown almost

Acknowledgement. This work was partially funded by the Bundesministerium für Bildung und Forschung of the Federal Republic of Germany (Förderkennzeichen 01AK045). The author alone is responsible for the content of the paper.

G. Carle and M. Zitterbart (Eds.): PfHSN 2002, LNCS 2334, pp. 229–242, 2002.

explosively throughout the last decade and some predictions say that the traffic growth may continue at well beyond a factor of two per year for at least a few more years [5].

Basic internet principles and mechanisms are significantly simpler and easier to operate than ATM, since the Internet Protocol (IP) [16] based networking concept does not need a path infrastructure and offers a unified 'best effort' type of service only. IP based packet switching networks have been installed in parallel to the Public Switched Telephone Network (PSTN) and they have already outgrown the PSTN in transfer capacity. Having done the exercise with ATM, it was obvious that IP based packet technology could as well be used for carrying voice, video and any other kind of electronically conveyable communications. The resulting new type of network which converges the services and applications of the ancestor networks on a single IP based platform and is open for a variety of new services and applications, is often refered to as 'Next Generation Network' (NGN).

This paper approaches NGNs from the service and application point of view and derives from that the capabilities and properties expected from an NGN as an evolution towards high speed carrier networks. The key requirements from the networking point of view are discussed and an outline of an NGN architecture is presented. Finally a few guidelines for a possible solution approach are developed.

2 Next Generation Networks

2.1 Roots and Standardization Status

Just ten years ago telecommunication service providers and carriers were mainly focusing on telephony and possible evolutions of telephony-like services towards higher value audio, video and multimedia applications. Around the same time the general availability of personal computers (PCs) as affordable tools for desktop computing for business and home applications had created a demand for simple, cost efficient and easy to use data communications and information access. Internet technologies and related application services opened the gates for PCs and Local Area Networks (LANs) towards universal and ubiquitous information exchange.

Technology evolution has turned PCs into high performance terminals and the affordability of broadband access even for residential users via xDSL and cable modems fuels the demand for bandwidth, throughput and performance of the network. IP packet forwarding has moved from general-purpose processor software to dedicated hardware devices and has made state-of-the-art routers capable of serving high-speed interconnection links at wire speed possible. Such progress allows a steadily increasing variety of different new services and applications to be created and delivered via internet technology based infrastructures. The delivery of (voice) telephony services over IP based networks including the traditional 'best effort' internet has already become a reality. New dedicated NGN solutions will provide the capabilities, features and functions required for the deployment of high speed, high performance telecommunications and data services with true 'carrier grade' service characteristics.

The Internet Engineering Task Force (IETF) [6] has proposed mechanisms like Integrated Services (IntServ) [7] and Differentiated Services (DiffServ) [8] for reservation of resources and differentiation of individual traffic flows in routers in order to facilitate service delivery with different levels of Quality of Service (QoS) in IP networks. Special protocols like RTP and RTCP [9], [10] have been designed to support and control the transport of real time applications as for example voice, audio or video communications across IP based networks. These are just a very few examples out of a variety of mechanisms and protocols that are discussed and proposed by IETF and many other standardization organizations as possible building blocks for a global set of NGN standards.

The most comprehensive framework has been initiated by the European Telecommunication Standards Institute (ETSI) with their TIPHON project [11]. Although initially mainly interested in the delivery of (voice) telephony services over IP based networks and the related interoperability issues between traditional telecommunications and IP worlds, TIPHON has subsequently broadened its charter towards more general issues of heterogeneous and multi-service packet networks. One of the key objectives of TIPHON's recently started NGN activity is a global consolidation and harmonization of NGN standardization in partnership with other organizations working in this field.   The latest version of TIPHON documents (Release 3) is publicly available at [12].

2.2   Services and Applications

NGNs are intended to accommodate and facilitate the widespread deployment of 'classical' and future telecommunication services and applications together with 'traditional' and new internet services on a common IP based networking platform. A short look at some major differences in the type and nature of some possible services will indicate the size of the challenge.

Traditional internet services and applications are usually based on direct host to host (or host to server) communications. The role of the Internet Service Provider (ISP) is mainly restricted to subscriber authentication, authorization and accounting (AAA) and the provisioning of access to the network. All service/application related features and functions are completely hosted in and initiated from the application terminal (host). Applications are adapted to operate in an uncontrolled, resource shared and 'best effort' only network environment by the Transmission Control Protocol (TCP) [13], which takes care of reliable transport as well as adequate utilization and fair sharing of available network resources. Data throughput and delay for communications are unpredictable and the resource sharing can cause strong interdependencies between simultaneously active communications.

Classical telecommunication services usually employ service control instances provided by a service and/or network provider. These service control instances take care for the availability of network resources, e.g. a connection path, but they also may offer a variety of additional capabilities, features and functions to be requested by a user application (e.g. ISDN supplementary services). Communications run on dedicated paths using exclusively reserved resources and the transmission behaviour in terms of throughput and delay etc. is predictable and 'guaranteed'. Consequently, there is no interdependency between different communications as long as the network design and dimensioning has been done properly. A major difference compared to internet services is that telecommunication service users traditionally expect a much better availability and reliability of their services.

The future of internet services will definitely include a more efficient usage from the application point of view (e.g. high-speed web surfing). Built on today's mechanisms this will definitely require a much faster (broadband) network. Another approach aims at a differentiation of service levels, e.g. to distinguish between gold, silver and bronze service. Future services and applications, still unknown yet, may raise additional requirements in terms of throughput, delay, reliability, security or whatever else. Evolution of telecommunication services still follows the vision of broadband real time services for true interactive (dialogue) communications, e.g. high quality audio/video communications up to the notion of 'virtual presence' (or 'telepresence'), in a fully multimedia enabled environment.

Converging all these services onto a common IP packet based NGN platform definitely requires a high speed QoS enabled network, i.e. a network with capabilities to deal with high volumes of data within well defined and distinct limitations on transfer delay and probability of loss. The network architecture has to be open and sufficiently flexible and scalable in order to accommodate new services and their specific requirements.

Flexibility and scalability are specifically important from the network control point of view. Different from telecommunication services, which usually create single, long duration data flows, many internet applications are composed from a multitude of short communication relations with different partner entities. This kind of behaviour may have to be extrapolated into future, yet unknown, NGN services with QoS requirements above the best effort service level.

2.3   Transport Protocols

Two different transport protocols are used with traditional internet applications: the Transmission Control Protocol (TCP) [13] and the User Datagram Protocol (UDP) [14]. UDP offers a completely connectionless transport of individual data segments (datagrams) through IP networks, whereas TCP provides a reliable transfer of contiguous data, i.e. the notion of data streams, over IP based packet networks. A good overview on the current status of these protocols and their related mechanisms is given in [15]. The recently developed Stream Control Transmission Protocol (SCTP) [16] is intended for message-oriented applications, e.g. reliable transportation of signaling data.

TCP's emphasis is on reliable delivery of data even in case of adverse terminal or network conditions. For that purpose it offers specific flow control and congestion avoidance mechanisms which have been refined and improved over several generations of TCP/IP implementations. The basic mechanism behind TCP is its acknowledgement controlled sliding window based data transfer. This kind of mechansim works well as long as the relations between the triple of expected throughput, window size and data-round-trip-delay can be kept within certain reasonable ranges and network and terminal buffers are well dimensioned. For high speed data transfer over wide area networks the inevitable physical propagation delay, which is ruled by the speed of light, may become a dominating parameter. Simply increasing the window size may affect the efficiency of flow control and congestion avoidance and jeopardize the objective of fair sharing of network resources. A careful tuning of window sizes and case by case selection of supporting mechanisms may improve, but probably not always solve the problem. Still, if there is no better alternative, the long time proven TCP will remain the protocol of choice.

As TCP aims for reliable delivery of data, even at the expense of delay (retransmission), it is not well suited for the transport of real-time applications. Congestion avoidance mechanisms, which are very useful in a resource shared, 'best effort' oriented environment, may be in contradiction to the target of sustained throughput with agreed QoS levels. The Real-Time Transport Protocol RTP [9], [10] has been specified as a mechanism to support end-to-end delivery of information with real-time characteristics in single ended as well as multicast applications and it may as well be applied for other QoS dependent services. RTP includes no flow control and since it does not include all necessary transport layer functions, it 'borrows' missing functions from an underlying transport protocol, which is usually UDP. RTP supports the applications with timing, sequencing, monitoring and other functions, but it does not provide any mechanisms to ensure timely delivery of data nor does it provide any means to guarantee delivery of data or a certain QoS. Lower layers are expected to provide suitable mechanisms to ensure these capabilities.

Since many of the future services and applications to be supported by NGNs are still unknown, it cannot be stated today whether the available set of transport protocols will be sufficient on the long run to link all NGN services and applications to the network.

2.4   Interoperability with Existing Networks

NGNs will provide a plenty of new services and applications on a new, special feature enabled, packet based networking platform. NGNs also will converge the full spectrum of already existing telecommunications and internet services on this same platform. A smooth introduction of NGN technologies will be based on a long-term coexistence of NGNs with traditional networking technologies.

This may be less critical with regard to the Internet since both networks are built from the same basic principles and mechanisms and their technological affinity will ease seamless interoperation. Needless to say, that interworking will only be possible on the basis of services and features that are supported by both technologies. The very fast growth of the internet has created a phenomenon never known by traditional telecommunication service providers and carriers. Equipment or even equipment family turnaround times down to less than three years have been reported from ISPs and IP network providers. Together with the desirability of 'better service' for end users and expected reductions in operational cost, this fuels the assumption that NGN technology, if matured and available in time, could potentially absorb and replace the traditional internet within less than a decade.

In the telecommunications arena the situation is somewhat different. A huge base of telephone network equipment is installed and operating satisfactorily in the PSTN. As long as the operation of this equipment is economically justifyable, (in other words, as long as there is a sufficiently high number of subscribers satisfied with the services and applications offered by this technology,) there is no reason and no pressure to remove or replace this equipment. Finally, the success and the speed of deployment and dispersion of NGN services will play a decisive role.

Interworking with the PSTN will be based on telephony gateways, which are capable to distinguish and interwork voice telephony, fax and low speed dial-in internet access services. Conversion between TDM and IP packet based transport will have to be provided and a lot of peering issues, starting from proper QoS mapping up to tarifing and billing issues, will have to be solved. Realistically, a survival time of PSTN equipment of at least several decades has to be assumed.

3   Key Requirements on Next Generation Networks

In this section some key requirements on NGNs as a new generation of carrier networks are discussed. The selection is driven by those aspects, which are important for their classification as 'carrier grade'[1] networks.

3.1   Quality of Service (QoS)

QoS is always the first (and sometimes the only) requirement that pops up in discussions about multi-service packet networks. A quick and easy answer to the problem could be to dimension and operate the network at a unified service level that satisfies the requirements of the most demanding application. However, an economically justifiable network operation will require a differentiated treatment of the variety of services and applications according to their specific needs.

QoS mechanisms in packet networks have to respect the characteristics of the different traffic flows in terms of their variance or even a more pronounced burstiness up to the extremes of a direct or overlaid on-off behaviour. Statistical methods have to be applied in order to describe such kinds of traffic, to analyze their interactions in the network, and to understand the resulting effects in terms of throughput, delay and loss of packets. Finally, the capabilities of the applications to tolerate (or not) a certain level of impairments induced by the network determine their requirements.

---

[1] The term 'carrier grade' probably has its origin in the high speed router start-up scene, where it was used to indicate that a planned router product was intended to provide the same level of service, the same level of serviceability and all the other nice features and properties, that people were used to find in the well established and mature PSTN backbone switching technology of established (long distance) carriers.

As an example, a measure for a very demanding service can probably be derived from the already mentioned audio/videocommunication with the notion of 'virtual presence'. High quality, high resolution video requires sufficient throughput and low loss, real-time interactive human (dialogue) communications cannot tolerate too much delay. Intermediate levels could be marked by voice telephony, which can tolerate some packet loss as long as certain delay limits are not exceeded, and a privileged internet service, e.g. with guaranteed minimum throughput. At the low end are traditional best effort internet services.

Since the provision of individual QoS levels per application instance or individual data flow is not feasible, requirements are usually mapped to predefined network services (or traffic classes), that provide a certain well defined and 'guaranteed' level of QoS. Requirements not directly matching with a network service then have to be mapped to the next better one. Administration, operation and supervision of network services and especially the process of assigning network services to different data flows may still turn out to be quite complex and expensive. Therefore a proliferation of network services has to be avoided. A low single digit number is recommended.

QoS can be measured at the technical level in terms of throughput, packet loss and packet delay, and these are the parameters that usually are influenced (but not exclusively determined) by the network. The decisive judgment criterion, however, will always be the user's perception. The network has done a good job as long as the user has a working application and the impression of a good service. Doing more than necessary usually causes unnecessary cost. Therefore a good understanding of applications and their capabilities is inevitable for a good network design and proper network and traffic engineering.

Another important aspect of QoS is its need for control. QoS requires resources to be available at the place and time where and when they are needed. A proper allocation has to be done wherever resources are limited. Since network control performance is an important cost factor, a good network design has to reflect the impacts of resource control, application behaviour (e.g. single or multiple flows) and related usage patterns (i.e. frequency and duration of usage).

3.2 Resilience

Network resilience describes a network's capabilities to provide sustainable service at agreed QoS levels under varying traffic conditions and in spite of different kinds of impairments affecting it. Such impairments may be caused by network internal or external events and appear as temporary or local overload, unavailability of certain network resources or any other effects.

Circuit switching based telecommunication networks provide up to more than five nines (99.999%) of service availability, a level far beyond that of many of today's data networks. This difference is mainly based on much more local redundancy, intrinsically fault tolerant network nodes and faster fault recovery mechanisms combined with a higher stability of software deliveries. Packet networks may even achieve a higher overall survivability, but currently available mechanisms are comparably slow.

Telecommunication subscribers are used to such high levels of service availability supported by fault recovery mechanisms that in many cases leave no perceptible or at least no annoying impairments on their delivered services (respectively the related QoS) and this experience determines user expectations for existing and new services in NGNs. Substancial efforts will have to be made in order to provide mechanisms efficient and fast enough to comply with these expectations without giving up the advantages in flexibility, simplicity of operation and operational cost currently fueling the trend for a transition towards packet based NGN solutions.

3.3 Security

Reliable network operation is heavily related to network security. Network elements have to be protected against any kind of unauthorized access. Malicious intruders might attack network control information and disturb network control communications and network operation up to complete network failure. Intruders could steal or modify administrative or operational data, e.g. subscriber profiles or charging data record (CDR) information. Intruders could intercept user traffic and violate subscriber privacy.[2] They might even modify or destroy user traffic data. The unauthorized access issue is even more critical in IP based networks compared to traditional circuit switched ones, since in most cases network control communications will use the same mechanisms and even share network resources with user traffic. Special care will have to be taken on these aspects.

A related issue is unauthorized usage of network resources. Since all QoS agreements rely on the (controlled) availability of required network resources, any unexpected traffic might significantly impair the QoS of regular traffic. Special attention has to be paid to intentionally malicious user traffic that could aim at denial of service or at other impacts on traffic handling or traffic control (including signaling) related entities.

3.4 Scalability

Another aspect with potentially significant impact on network economics is scalability. NGNs should match the needs of small local operators as well as those of large international carriers. Network equipment should enable a seamless and ideally linear adaptation to increasing numbers of subscribers as well as changes in service utilization and traffic patterns. It has to be open to accommodate different requirements of new emerging services and applications. Scalability should cover all network components, capabilities and functions for simultaneous as well as independent adaptation.

---

[2] Note, that on the other hand lawful interception must be supported.

3.5   Serviceability

High reliability and service availability require a seamless operation of all vital network functions. Maintenance and service activities on network components should not impact the perceived functionality, quality and performance of user applications. Addition or change of equipment and introduction of new software releases for capacity, performance and/or functional upgrades should be possible during normal system operation and without traffic interruption. Modular systems should provide for independent maintenance of different modules and functions. Such requirements request a careful reflection in the architecture and design of network and equipment and their way of operation.

3.6   Economy

Many of different factors contribute to the overall economics of owning and operating a network. QoS, reliability and security contribute to initial procurement and depreciation as well as to any subsequent expenses. Further expenses for extensions and upgrades are heavily influenced by scalability and serviceability. The major cost contribution for a long-term ownership, however, may be accumulated from day-to-day operation and maintenance. Reduction of operational cost is propagated as a main driving force for convergence to IP based NGNs, and the reasoning behind is simplicity and ease of operation. To preserve this paradigm, significant attention should be focused on related issues through all phases of NGN lifetime starting with initial concepts.

4   Next Generation Network Architecture

NGN architecture is driven by several factors. They all end in the target of a converged, unified networking platform open for and capable of supporting a variety of different services and applications. Still, the network of the future will be much less unified than the incumbent PSTN. Telecommunication deregulation breaks monopolies and fuels competition and differentiation among carriers (network operators) and service providers. The resulting network (Fig. 1) is composed of several (NGN) network domains, which may be owned by different operators or service providers. Application services, either hosted in user terminals (internet model) or provided and controlled by service providers (telecommunications model), compete for transport resources provided and controlled by the network domains.

A clear separation between network control and service control with well-defined interface and protocol standards is required to make this model work. As a result, the

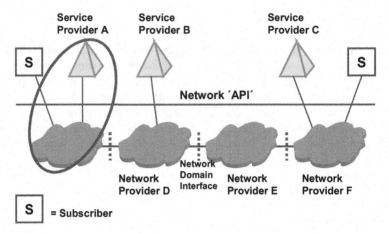

Fig. 1. A generic NGN architecture. Note, that Service Provider A also acts as Network Operator

network will form the envisaged universal platform providing some kind of a generic 'Application Programming Interface' (API) for the support of new services.

Network operators may employ different mechanism to ensure QoS, resilience and security within their different network domains. As a consequence, their definitions and implementations of network services (traffic classes) may differ and the probability to reach a consensus for a globally agreed and standardized set of 'well known network services' (that was achieved with ATM) may not be very high. Thus, the interface standards for network domain interoperation have to include mechanisms capable to cope with differing network service specifications. As an example, explicit QoS requirements could be signaled and the mapping to suitable network services could be done within the domains (Fig. 2).

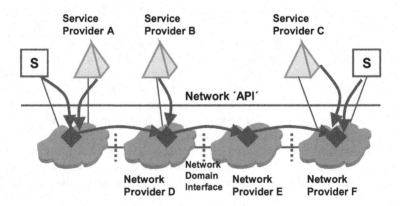

Fig. 2. QoS signaling relationships in a generic NGN architecture

Different applications may use different transport protocols and the different network domains may employ completely different lower layer mechanisms in order to achieve the required properties. Finally, the only remaining common denominator is the networking concept based on the Internet Protocol (IP) [16] (Fig. 3).

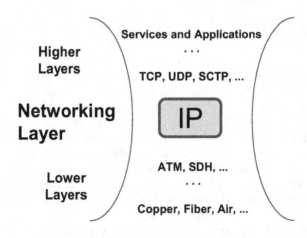

Fig. 3. Internet Protocol (IP) based networking is the common denominator of NGNs

5   A Few Considerations towards an Implementation

5.1   What Is Available

A good solution approach should fulfill all requirements, provide all the features and adhere to all the promises and expectations that have been built around the NGN idea. If we measure the hottest candidates of todays discussions against this objective, it may turn out as follows:

The RSVP based Integrated Services approach [7] succeeds with QoS, but at the expense of a very sophisticated and complex reservation procedure. There seems to be a wide consensus that RSVP does not scale for large networks or frequently changing connection environments. Besides, IntServ does not address network resilience issues.

The Differentiated Services approach [8] is comparably simple and therefore it can scale. However, DiffServ does only provide a few elements of a complete QoS solution, i.e. some kind of an incomplete toolset (edge functions, per hop behaviours) for a differentiated treatment of traffic in a network. DiffServ also does not address network resilience issues.

A more comprehensive approach is Multiprotocol Label Switching (MPLS) [17], [18]. MPLS addresses (to a different extent) traffic engineering, resource reservation, reliability, security and scalability issues. Like ATM, MPLS employs a path infrastructure, which is used to direct 'connection traffic' (or 'flows'), and like ATM, with every issue it tackles, it packs another piece of complexity on its back. Finally, somebody will have to raise the question: isn't it just ATM - in a different colour?

5.2 The Case for a 'Stateless' Core

Traffic management with a path and connection based concept means complexity, complexity for setting up and maintaining the path infrastructure as well as for managing the resources within the different paths on each individual link along each path. This complexity requires the intelligence of well educated staff and in an environment with highly volatile connection traffic it requires permanent staff attention. Permanent attention of well educated staff is very expensive.

High reliability in a dedicated path concept means redundancy, which has to be provided, dedicated and reserved separately. This redundancy again has its cost. High network resilience requires fault tolerance, i.e. the capability to provide an adequate replacement of lost resources within a very short time in case of failures, so that the applications are not (or only marginally) affected. For that purpose either redundant paths have to be available and (pre)configured in parallel to the actually used paths (hot standby) or the control system has to be able to take care of rerouting, i.e. setting up of new paths and reconfiguration of all afflicted connections, within a very short time. This again may turn out to be quite costly. Additional effort may be required for adaptation of the path infrastructure in case of network extensions or with upgrades, etc..

The key issue with path and connection based concepts is the use of 'states' for each path and each connection in all network nodes and on all links, wherever resources may be shared between different paths and/or connections. States are managed through information exchange between related entities and they are manifested in the different entities as specific data sets or table entries. Every change related to a path or a connection results in at least one 'state change' in the network control system. It has been recognized as one of the key advantages of the Internet concept that it operates in a 'stateless', connectionless way.

5.3 A Proposal for a Connectionless NGN Domain

To take full advantage of the capabilities of IP based packet data transport it is proposed to limit any notion of connections or predefined paths to the borders of NGN domains. As a result, the core of the domain will operate in a 'stateless', connectionless manner. For the provision of QoS the overall capacity budget of the network is calculated and related shares are derived and allocated to the network borders. Network Admission Control (NAC) is done on both ingress and egress borders, in order to avoid egress congestion. Best effort traffic may run uncontrolled, because it can be pushed out by QoS traffic in case of congestion.

To prevent traffic congestion or hot spots inside the network, the network nodes are authorized to distribute the traffic autonomously over all reasonably useable paths towards the destination border. The distribution scheme may use a per packet or a (local) per flow paradigm. Reasonably useable paths may include all outgoing links that approach the destination border without creating loops and within the specified QoS boundaries (e.g. delay limits). The knowledge about reasonably useable paths may be derived from link state information as usually exchanged by routing protocols.

The autonomous, local distribution of traffic opens new possibilities for network resilience. Provided that fast fault detection mechanisms are applied, faulty parts (links or nodes) in the network can be isolated from traffic almost immediately by a local reaction of their neighbour nodes, which changes their traffic distribution patterns. If network admission control has been done carefully, the QoS specifications will still be fulfilled and the user may not even note the event.

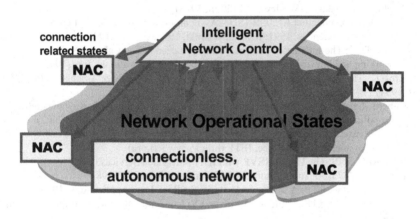

Fig. 4. Basic architecture of an autonomous, connectionless network domain

An intelligent control system can be put on top of this network domain. This control system takes the role of a network supervisor in terms of collecting information from the network and providing necessary corrective instructions, e.g. changes of strategies or parameters, to the NACs and the network nodes. Fig. 4 illustrates the basic architecture of this type of almost autonomous network domain, that should easily scale for new service and application demands (no user traffic related states in the core) and signifcantly reduce the operational cost for the operator (autonomous operation and adaptation to changing network/traffic conditions). For security and serviceability no extraordinary impacts are expected.

6 Conclusion

The Next Generation Network concept as a future high speed public network infrastructure has been introduced. Its origin and its operational environment have been described and the key requirements towards 'carrier grade' NGN solutions have been discussed. A related overall network architecture with its constraints and interfacing issues has been presented and some requirements for standardization have been derived. Finally, a proposal for a possible NGN domain architecture has been presented. The proposal, based on a connectionless IP core, appears quite promising with respect to the overall requirements. We are performing further work in order to elaborate and prove the feasibility of this approach.

References

1. Händel, R., Huber, M.N., Schröder, S.: ATM Networks. Concepts, Protocols, Applications. Addison-Wesley (3rd Edition, 1998)
2. ITU-T Recommendation I.150: B-ISDN Asynchronous Transfer Mode Functional Characteristics (Geneva, 1991; revised Helsinki, 1993; Geneva, 1995)
3. Comer, D. E: The Internet Book. Everything you need to know about computer networking and how the Internet works. Prentice Hall International, Inc. (1995)
4. Krol, E.: The Whole Internet User's Guide & Catalog. O'Reilly & Associates, Inc., 2nd Ed. (1994)
5. Roberts, Lawrence G.: Internet Traffic Measurement 2000 and 2001. Keynote address at Metropolitan Communications Conference, San Francisco (January, 16th, 2002) (http://www.cibcwm.com/conferences/metrocomm/)
6. Internet Engineering Task Force, IETF (http://www.ietf.org)
7. IETF RFC 2210: The Use of RSVP with IETF Integrated Services (1997)
8. IETF RFC 2475: An Architecture for Differentiated Services (1998)
9. IETF RFC 1889: RTP: A Transport Protocol for Real-Time Applications (1996)
10. IETF RFC 1890: RTP Profile for Audio and Video Conferences with Minimal Control (1996)
11. Telecommunications and Internet Protocol Harmonization Over Networks, TIPHON (http://www.etsi.org/tiphon)
12. ETSI TIPHON Documentation: Release 3 (http://portal.etsi.org/tiphon/Marketing_Release3.asp)
13. There is a plenty of literature on TCP/IP and its recent implementations. The original document is IETF RFC 793: Transmission Control Protocol (1981)
14. IETF RFC 768: User Datagram Protocol (1980)
15. Hall, E.A.: Internet Core Protocols: The Definitive Guide. O'Reilly & Associates, Inc. (2000)
16. The original document on the Internet Protocol is IETF RFC 791: Internet Protocol (1981). A good summary on the recent status is given in [15].
17. IETF RFC 2702: Requirements for Traffic Engineering over MPLS (1999)
18. Davie, B., Rekhter, Y.: MPLS. Technology and Applications. Academic Press, San Diego (2000)

Protocols for High-Speed Networks:
A Brief Retrospective Survey
of High-Speed Networking Research

James P.G. Sterbenz

BBN Technologies, 10 Moulton St., Cambridge, MA 02138-1191 USA
jpgs@ieee.org
http://www.ir.bbn.com/~jpgs

Abstract. This paper considers high-speed networking research from a historical perspective, and in the context of the development of networks. A set of axioms guiding high-speed network research and design are first presented: Ø Know the Past; I Application Primacy; II High Performance Paths; III Limiting Constraints; IV Systemic Optimisation. A framework of network generations is used as the basis for the historical development of high-speed networking: 1st – Emergence; 2nd – Internet; 3rd – Convergence and the Web; 4th – Scale, Ubiquity, and Mobility. Each generation is described in terms of its application drivers, and important infrastructure and architectural characteristics. Woven into this historical thread are the important research thrusts and sub-disciplines of high-speed networking, and their impact on deployment of the Global Information Infrastructure. Based on this historical perspective, a set of Systemic Optimisation Principles are identified: 1 Selective Optimisation; 2 Resource Tradeoffs; 3 End-to-End Arguments; 4 Protocol Layering; 5 State Management; 6 Control Mechanism Latency; 7 Distributed Data; 8 Protocol Data Units. We are now in the state where everything has some aspect of high speed networking, and nothing is only about high-speed networking. This is a double-edged sword — while it reflects the maturity of the discipline, it also means that very few people are looking after the performance of the entire Internet as a *system of systems*. Rather, performance analysis tends to be isolated to individual network components, protocols, or applications. Furthermore, the high-speed networking community is not pushing back at the multitude of deployment hacks by network and application service providers (such as middleboxes) without regard to global network performance effects. Thus, this paper argues that the high-speed networking community should have the future role of caring about high-speed network deployment on a *global* scale, and throughout the entire protocol stack from layers 1 through 7.

1 Introduction

Over the last twenty years or so, the discipline of high-speed networking has seen an emergence, significant activity, and melding into the mainstream of network research as a mature field. This paper aims to survey some of the most significant thrusts of high-speed networking in a historical context.

G. Carle and M. Zitterbart (Eds.): PfHSN 2002, LNCS 2334, pp. 243–265, 2002.
© Springer-Verlag Berlin Heidelberg 2002

High-speed networking is difficult to define, because what constitutes "high-speed" changes with time, as technology progresses and applications develop. Over time, the switching rates of electronic and photonic components increase, and the density of VLSI chips and optical components increase. This results in higher available bandwidths (data rates), increased processing capabilities, and larger memories available in the network and in end systems. Furthermore, the effective rate at which network components operate decreases as we move up the protocol stack and from the network core out to the end system. There are two reasons for this. The need for the network to aggregate vast numbers of high-performance interapplication flows dictates that the core of the network must be higher capacity than the end-systems. Furthermore, it is easier to design components for high-speed communication whose sole purpose is networking, than it is to optimise end systems with multiple roles and applications.

In the late 1990s, deployed link layers and multiplexors (such SONET) operated on the order of tens of Gb/s, switches and routers at several Gb/s *per* link, end system host–network interfaces in the range of 10–100 Mb/s, and applications typically on the order of several Mb/s. By the early 2000s, link bandwidth and switch bandwidth had increased an order of magnitude or two, but local access and end system bandwidth continued to lag.

High-speed networking consists not only of the quest for high bandwidth, but also for low latency (or the perception thereof) and in the ability to cope with high bandwidth-×-delay product paths; these will be motivated in the next section.

This paper draws heavily and quotes from two earlier works by the same author. The historical framework as a series of networking generations was introduced in 1994 at *Protocols for High Speed Networks* in [25], and later updated by [26] with the addition of a fourth generation. The high-speed networking axioms and principles were developed for [26], from which the figures and much of the text in Section 3 is derived.

The rest of this paper is organised as follows: Section 2 introduces a set of axioms to guide and motivate high-speed networking research. Section 3 presents a historical view of networking as a sequence of generations. Into this generational perspective are woven some of the most important research pursuits within the high-speed networking discipline, as are a set of high-speed networking principles. While a few references to the literature are provided, [26] should be consulted for a significantly more comprehensive and complete bibliography. Section 4 considers the future role of high-speed networking research.

2 Axioms for High-Speed Networking Research

High-speed networking is a mature discipline, but there has been little attempt to structure and document the axioms and principles that have guided high-speed networking research and system design. In this section a guiding set of axioms are presented (quoted from [26]):

Ø. KNOW THE PAST, PRESENT, AND FUTURE: *Genuinely new ideas are extremely rare. Almost every "new" idea has a past full of lessons that can either be learned or ignored. "Old" ideas look different in the present because the context in which they have reappeared is different. Understanding the difference tells us which lessons to learn from the past and which to ignore. The future hasn't happened yet, and is guaranteed to contain at least one completely unexpected discovery that changes everything.*

I. **APPLICATION PRIMACY:** *The sole and entire point of building a high-performance network infrastructure is to support the distributed applications that need it. Interapplication delay drives the need for high-bandwidth low-latency networks.*

This principle is the motivation for why we need high-speed networking; if inter-application delay is low enough, there is no difference to the user between a centralised application and one that is distributed across the globe. End-to-end latency must clearly be low enough, and bandwidth must be high enough that the transmission delay of data (first bit to last bit) is small enough.

II. **HIGH-PERFORMANCE PATHS GOAL:** *The network and end systems must provide a low-latency high-bandwidth path between applications to support low interapplication delay.*

III. **LIMITING CONSTRAINTS:** *Real-world constraints make it difficult to provide high-performance paths to applications.*

These constraints include the speed of light, limits on channel capacity and switching rate, heterogeneity, policy and administration, cost and feasibility, backward compatibility, and standards. It is important to attempt to distinguish reasonable constraints from those that do not have sound basis; this will be reconsidered in Section 4.

IV. **SYSTEMIC OPTIMISATION PRINCIPLE:** *Networks are* systems of systems *with complex compositions and interactions at multiple levels of hardware and software. These pieces must be analysed and optimised in concert with one another.*

It is this axiom that can be refined into a number of high-speed networking principles. In the next section, these refinements will be introduced in the historical context of high-speed networking research.

3   A Brief History of High-Speed Network Research

This section traces the intertwined history of network development with high-speed networking research. The history of networking can be divided into generations [25, 26] that capture significantly different characteristics in their development and deployment. Coincidentally, these generations correspond roughly to the four decades since the 1970s.

3.1   First Generation – Emergence

The first generation lasted through roughly the 1970s and is characterised by three distinct categories: voice communication, broadcast entertainment, and data networking, each of which was carried by a different infrastructure. Voice communication was either analog circuit switched over copper wire (and microwave relays) in the public switched telephone network (PSTN), or free space analog radio transmission between transceivers. Entertainment broadcasts to radio receivers and televisions were carried by free space broadcast of RF (radio frequency) transmissions. There was little need for high-speed networking since all of these applications had well-defined, and relatively low bandwidth requirements.

Data communications was the latest entrant, and provided only a means to connect terminals to a host. This was accomplished either by serial link local communications (*e.g.* RS-232 or Binary Synchronous Communications used on mainframes), or by modem connections over telephone lines for remote access; in both cases copper wire was the physical medium. Packet networking began to emerge in the wired network in ARPANET and packet radio, primarily for military applications [3]. Early LAN research led to the emergence of Ethernet [15] and token ring, which used a shared medium for communication among multiple end systems.

Early packet routers used a bus-based general-purpose computer system with multiple network interfaces (NIs), which stored each packet in main memory to be forwarded out the appropriate interface after network layer protocol processing was performed. This architecture is shown in Figure 1.

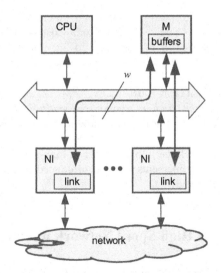

Fig. 1. First Generation Store-and-Forward IP Router

During these formative years of data networking, the discipline of high-speed networking didn't exist as such, but rather performance was one of many considerations that was brought to bear as necessary for network research and development. Store-and-forward message and packet switching and shared medium LANs were generally not the limiting performance factor given the computing capabilities of end systems and the link technologies of the 1970s.

3.2 Second Generation – the Internet

In roughly the 1980s a dramatic jump in the types and scope of networking occurred, but the three categories of communication (voice, entertainment, and data) remained relatively distinct. This period took us from the experimental ARPANET to the ubiquitous Internet.

While the end user of the voice network generally continued to use analog telephone sets, the internal network switches and trunks became largely digital, but transmission remained mostly over copper wire and microwave relay links. Additionally, there was widespread deployment of digital PBXs (private branch exchange telephone switches) on large customer premises. Mobile communications began to emerge in form of cellular telephony. The significant addition to the entertainment category of networking was the wide scale deployment of cable television (CATV) networks for entertainment video over copper coaxial cable. So while the PSTN evolved internally and CATV infrastructure was built, the application requirements saw little change. The impetus for high-speed networking consisted primarily of increasing aggregate link and switching capacity deployed deep within the PSTN.

In data networking, we first saw the emergence of consumer access, but in the primitive form of bulletin board systems (BBSs) and consumer online services (such as America Online, CompuServe, and Prodigy). These were essentially first generation networks made available to consumers, with modems connecting to a central server farm, and there was little impact on the emerging Internet.

Connection oriented corporate enterprise networks using protocols such as BNA, DECNET, and SNA were widely deployed, along with the deployment of public X.25 networks (used primarily as corporate virtual private networks). Most of these networks used copper wire as the predominant physical medium. These networks used incompatible architectures that were poorly interconnected with one another, if at all. While there were significant research advances in the context of enterprise networks in the second generation, there was little direct impact on the emerging Internet.

The collection of research and education networks such as BITNET, CSNET, and UUNET were collectively referred to as the Matrix [21] before they began to join the Internet, unified by IP addresses, with DNS symbolic addressing replacing bang paths. The growth in data networking for universities and the research community was significant during this period, for purposes of file transfer, remote login, electronic mail, and Usenet news. The technology employed was the packet switched Internet, utilising the TCP/IP protocol suite. In the wide area, the backbone network consisted of store-and-forward routers connected by leased 56kb/s telephone lines. The NSFNET upgraded the infrastructure to 1.5 Mb/s T1 lines (and ultimately 45Mb/s T3 lines at the transition into the third generation). In the local area, shared media Ethernet and token ring networks became ubiquitous and allowed clusters of workstations and PCs to network with file and compute servers.

The second generation is the time in which high-speed networking came into existence as a distinct discipline. As the Internet and enterprise networks came into wide use by the research and business communities, respectively, there was a corresponding desire to support more sophisticated applications with better performance.   Remote terminal access was significantly slower than local connections, while file and document transfers took longer than users desired.   It was clear that higher speed networks would enhance productivity.   Similarly, it was recognised that significantly higher aggregate bandwidth was needed to satisfy the increasing demand for network services (as in the case of the PSTN).   This drove the Internet link bandwidth mentioned above, enabling more users to put greater demand on the network.   The cycle of demand and capacity increase had begun, and high-speed networking research rose to meet the challenge, initially at the lower layers of the protocol stack.

Network. By the mid 1980s, the Internet was seeing significant growth, and shared medium LANs were reaching capacity, requiring the deployment of bridges to increase spatial reuse, and demanding faster link rates (*e.g.* from 4Mb/s to 16Mb/s token ring, and driving research into technologies such as FDDI).   The goal of the high-speed networking community was to increase network link bandwidth by a couple orders of magnitude beyond that supported by the store-and-forward IP routers and deployed shared medium LANs, to gigabit per second data rates.   The initial steps were the SONET OC-3 (155Mb/s) and OC-12 (622 Mb/s) rates.   Furthermore, there was the desire to support integrated networks, in which data, voice, and video could be carried on the same network.

Since conventional *per* packet datagram forwarding was too complicated to consider at these data rates in the technology of the time, fast packet switching was proposed (e.g [34]).   By substantially reducing the complexity of packet processing, hardware implementation of the switching function was possible.   There are four key motivations that drove fast packet switching:

1. Dramatically simplifying packet processing and forwarding lookups by establishing connection state.
2. Eliminating the store and forward latency.
3. Eliminating the contention of the general-purpose computer bus as the switching medium.
4. Adding the ability to provide QOS guarantees to emerging multimedia applications, facilitated by resources reservations for connections.

The architecture of a fast packet switch is shown in Figure 2.   The goal is to blast packets through the switch without the delays of blocking due to contention in the switch fabric or need for store-and-forward buffering.

PROTOCOL DATA UNIT PRINCIPLE: *The size and structure of protocol data units are critical to high-bandwidth low-latency communication.*

Fast packet switches are based on maintaining connection state to simplify the *per* packet processing as much as possible, so that the data path can be efficiently implemented in hardware.   This requires the latency of connection setup before data can be transferred; for long connections this cost is amortised over many packets, but the user and application requirements for fast connection setup were not very often considered, and there was very little emphasis on the high-speed implications of signaling and control.

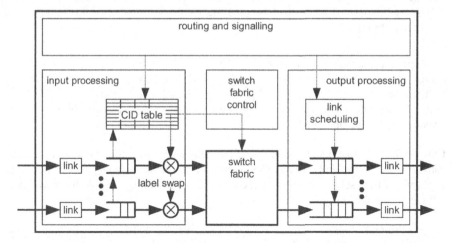

Fig. 2. Connection-Oriented Fast Packet Switch

CONTROL MECHANISM LATENCY PRINCIPLE: *Effective network control depends on the availability of accurate and current information.* Control mechanisms must operate within convergence bounds that are matched to the rate of change in the network, and latency bounds to provide low interapplication delay.

Further refinements of this principle to high-speed networking include minimising round trips for control messages, exploiting local knowledge, anticipation of future state, the proper balancing of open-and closed-loop control mechanisms, and the separation of distinct control mechanisms (such as error, flow, and congestion control). The existence of any state in the network requires its management, and difficult tradeoffs must be made:

STATE MANAGEMENT PRINCIPLE: *The mechanisms for installation and management of state should be carefully chosen to balance fast, approximate, and coarse-grained against slow, accurate, and fine-grained.*

In addition to enabling higher link rates, switched networks overcame many of the scalability limitation of shared medium networks, such as (the original) Ethernet and token ring. By using link protocols scalable in data rate (such as SONET), in conjunction with scalable switch architectures, networks could be easily grown in capacity, by increasing link rate and adding links, respectively.

The fast packet switching research took on a life of its own however, with the codification of ATM standards. Unfortunately, rather than migrating fast packet switching research technology into the Internet, an entire layer 3 routing, addressing, and signaling architecture was built for ATM, intended by its proponents to replace IP. And ATM was fraught with significant design flaws, from the small cell size that pushed the technology curve so far out that point-to-point 100Mb/s Ethernet chips became cheaper and more ubiquitous than ATM UNI chips, to complex traffic management and inefficient signaling messages. For better or worse, by the late 1980s, the IP based Internet had become the global information infrastructure, and any attempt to replace it was a futile exercise.

Transport Layer and End Systems. The late 1980s (into the early 1990s) saw intense research at the transport layer and in end system and host–network interface architecture (*e.g.* [7,9,23,33,36]. It also produced one of the most important principles to guide where functionality could, and should be placed, the *end-to-end arguments* [22], paraphrased here:

END-TO-END ARGUMENT: Functions required by communicating applications can be correctly and completely implemented only with the knowledge and help of the applications themselves. Providing these functions as features within the network is self is not possible.

This principle tells us that certain functions, such as end-to-end error control and encryption *must* be provided at the endpoints. Providing this functionality in the network does not preclude the need for end system implementation, and thus may be a waste of resources in the network.

END-TO-END PERFORMANCE ENHANCEMENT COROLLARY: *It may be useful to duplicate an end-to-end function hop-by-hop, if doing so results in an overall (end-to-end) improvement in performance.*

There are indeed justifications for a simple network that does not heavily rely on embedded stateful functions, including better resilience to link or node failures; this is one of the key ARPANET design decisions [14]. But the end-to-end arguments do not argue for a simple network *per se*. Rather the argument is that end-to-end functions should not be *redundantly* located in the network, but rather replicated where necessary to only to improve performance. This is a key principle in high-speed networking that indicates, for example, that hop-by-hop error control can shorten control loops such that end-to-end error control can be exerted less frequently with an overall reduction in latency to applications.

As deployed network bandwidth increased, and fast packet switch prototypes were built, it was recognised that the bottleneck in end-to-end communication was moving to the edges of the network. There was a period when the grand challenge of communications was to design networks capable of transferring data at rates in excess of 1 Gb/s. While fast packet switching research suggested that this was feasible in the network, delivering this bandwidth end-to-end was (and still is) more challenging. Protocol processing was constraining distributed processing, and it was commonly thought that the key bottleneck lay in the transport protocols. This resulted in significant debate between the advocates of new transport protocols, those who thought that protocols should be implemented in hardware in the network adapter, and those who thought that TCP would perform quite well if implemented properly. The following conjectures sumarise these positions:

EC1:   Designing a new transport protocol enables high-speed communication

EC2:   Implementing protocols on the host–network interface will enable high-speed networking

EC3:   Implementing protocol functionality in hardware speeds it up

While there is some basis for each of these statements, the mere replacement of an existing transport protocol (such as TCP) by a new transport protocol and implementing it in hardware on the host–network interface does not in itself solve the problem.

In the end, the transport protocol debate was irrelevant, due to the explosion of the Internet and pervasiveness of TCP. TCP is now the legacy data transport protocol of the global Internet, and for better or worse will be with us indefinitely. Thus the main thrust of research became how to optimise TCP for high performance given the evolution of high-speed network infrastructure, and what changes can be made in the protocol without breaking previous implementations [11].

It is critical to analyse existing end system architectures to determine where overhead and bottlenecks lie; this is the SYSTEMIC OPTIMISATION PRINCIPLE. It does little good to highly optimise operations that are not part of the bottleneck, or to create other bottlenecks as a side effect of an optimization, which leads to a particularly important refinement:

SELECTIVE OPTIMISATION PRINCIPLE: *It is neither practical nor feasible to optimise everything. Spend time and system cost on the most important constituents to performance.*

Considerations of the tradeoffs between hardware and software protocol functionality [4] and wide dissemination of the analysis of an existing protocol (TCP over IP) [5] provided needed perspective on where the bottlenecks really are, and what needed fixing. It was observed that the significant overheads were in the operating system and in *per*-byte operations such as checksumming and copying, as well as timer management. The approach shifted to systemic analysis and elimination of bottlenecks in the *critical path* with emphasis on related operating system and protocol implementation efficiencies, as well as providing sufficient memory bandwidth to the network interface [6,18,23]. These systematic analyses showed that areas to consider for reform included eliminating or reducing copies [23] and revisiting the I/O abstraction for communications [8,24]; that is, communication should be a first-class citizen, like memory or native graphics interfaces.

Protocol bypass [37] is a technique to optimise critical path processing, as shown in Figure 3.

The entire protocol stack is analysed to identify frequent operations, which are put in the bypass path, consisting of a single process without internal concurrency. A template is used to store packet header fields to quickly match or build headers (as in TCP header prediction). Data in the bypass path is shared with the conventional protocol stack. The templates are state that can be created by connection setup, or created dynamically in a data driven manner.

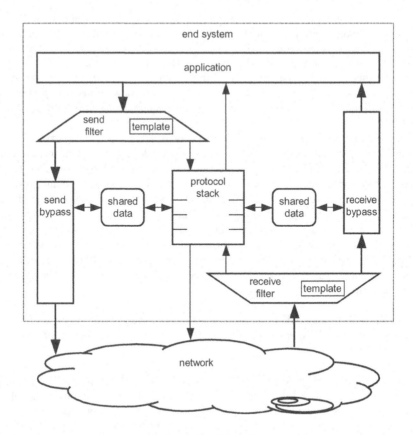

Fig. 3. Protocol Bypass

Many of the control overheads were a result of process *per* layer implementation of the protocol stack in conjunction with I/O mechanisms that were never designed for extremely high data rates, leading to the important principle that layering as an abstraction need not lead to layering as an implementation technique:

PROTOCOL LAYERING PRINCIPLE: *Layering is a useful abstraction for thinking about networking system architecture and for organizing protocols based on network structure. Layered protocol architecture should not be confused with inefficient layer implementation techniques.*

Integrated layer processing [6] is a way to overcome the overhead of layered system implementations, and can be viewed as the way to efficiently implement the bypass path described above. In a conventional layered protocol implementation, transport and network layer processing of data would consist of multiple distinct loops, as shown in Figure 4a.

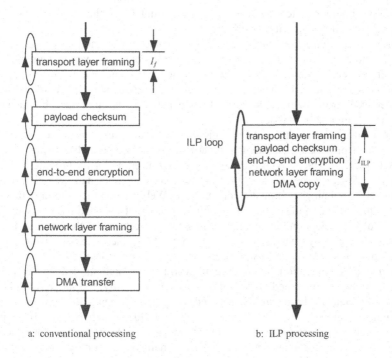

Fig. 4. Conventional and ILP Processing for TCP/IP

By employing ILP, all of the functions are processed in a single ILP code loop, as shown in Figure 4b. There are substantial savings in datapath processing by doing this. By leaving the data in place, copies between layers have been eliminated. Furthermore, joint code optimisations within a layer for *per* byte operations such as checksum and encryption may be possible. Additionally, by merging the processing loops for the various functions and putting them together, the overhead involved with transfer of control between the layers and functions is reduced. Hardware versions of ILP are also possible [23,10].

3.3 Third Generation – Convergence and the Web

The 1990s saw the emergence of integrated services: the merging of data, voice, and entertainment video on a single network infrastructure. With the advent of IP-telephony gateways, the PSTN started to become a subnet of the Internet, and with the advent of streaming multimedia, the same became imaginable for entertainment audio and video. Network service providers scrambled to keep capacity ahead of demand in over-provisioned networks, since the QOS mechanisms to support real-time and interactive applications were just beginning to emerge.

The second generation was characterised by the packet switched Internet, X.25, and enterprise networks. The third generation was characterised by an IP based global information infrastructure (GII) increasingly based on fast packet switching technology interconnected by fiber optic cable, and IP as the single network layer unifying previously disjoint networks.

The second significant characteristic of the third generation is in the scope of access, with consumers universally accessing the Internet with personal computers *via* Internet service providers (ISPs). Disconnected BBSs are a thing of the past, and online consumer services have become merely value-added versions of ISPs. The Internet went from being a kilobit kilonode network, through megabit meganode, approaching gigabit giganodes.

The final distinguishing characteristic of the third generation is the World Wide Web, which provided a common protocol infrastructure (HTTP), display language (HTML), and interface (Web browsers) to enable users to easily provide and access content. Web browsers became the way to access not only data in web pages, but images and streaming multimedia content. The Web became the killer app that drove bandwidth demand, and the rate of adoption of Internet connections vastly exceeded the rate of new telephone connections. In spite of the fact that the Internet and Web became the primary reason for users to have PCs, these devices were still not designed with networking as a significant architectural consideration.

In the third generation, high-speed networking research moved up the protocol stack to be more concerned with applications. Additionally, the failure of ATM and decreasing cost in hardware finally led to practical application of fast packet switching technology to IP routers in the late 1990s, which became IP switches. Optical networking saw some significant advances, which lead all but the most skeptical to consider that optical switching finally held some promise for future deployment.

This divergence of high-speed networking research into the upper and lower layers, respectively application layer and switch design, had the effect of fragmenting the discipline, and in mainstreaming high-speed networking into other sub-disciplines of communications, such as router/switch design and applications.

Internet. Demand for the Internet was steadily increasing by the end of the second generation, and short term solutions were necessary. This lead to a significant optimisation of IP router architecture to eliminate the shared CPU and memory as source of contention among the various network links. Distributing and offloading the network layer protocol processing to the NIs, as shown in Figure 5, accomplished this goal. This architecture was used in the NSFNET routers of the mid-1990s.

Packets are moved between NIs across the bus using *third party* transfers, without going through main memory. Each network interface contains a network interface processor (NIP), which performs the network layer processing, along with buffer memory for packet processing. While this significantly reduces the contention for a single memory and distributed the processing to each NI, this architecture still requires a store-and-forward hop on the NI. Furthermore, a single bus as the interconnect between all NIs significantly limits scalability.

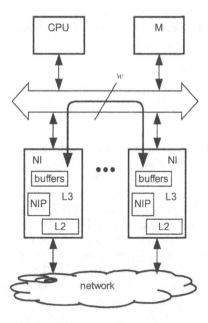

Fig. 5. Third Generation IP Router with Third-Party Bus Forwarding

Network. While research in fast packet switching had flourished in the second generation, this translated to only limited deployment in the third generation. ATM switches were deployed, but generally as islands of PVC meshes under IP. Performance was not particularly good, initially due to the inexcusable assumption that traffic was Poisson, and even after reasonably deep switch buffers were added, to incompatible forwarding and signaling mechanisms. While there were some hopes [25] and attempts (such as I-PNNI) to unify the IP and ATM frameworks, very little progress was made.

Two forces resisted the global deployment of a connection-oriented network layer, such as ATM. First, the explosion of the IP based Internet and Web in the mid 1990s entrenched TCP as the end-to-end protocol, and IP as the single global network layer; the *hourglass principle* indicates that there should only be one network layer. In the cases where connection oriented network protocols were deployed in backbone networks (such as ATM or X.25), IP traffic was run over these other network layers in a kludge of inefficient layering and incompatible control mechanisms that resulted in the native network layer being used as if it were a link layer. In the end, there was little motivation to create native ATM applications and transport protocols, or to use the ISO application protocols such as FTAM (file transfer, access, and management) or VT (virtual terminal).

Second, the limitations of shared medium link protocols such as Ethernet and token ring were overcome by the evolution of Ethernet to a switched point-to-point link protocol, with order-of-magnitude increases in data rate. This further reduced the motivation for adoption of ATM using scalable SONET links to increase the bandwidth on network links.

Finally, the dramatically increasing capabilities of VLSI processing in the 1990s finally made it feasible to consider *per* packet datagram forwarding and *per* flow queueing in switches.

RESOURCE TRADEOFF PRINCIPLE: *Networks are collections of resources. The relative composition of these resources must be balanced to optimise cost and performance. This relative cost changes over time, due to nonuniform advances in different aspects of technology.*

Therefore, much of the research community shifted their attention to speeding up connectionless datagram forwarding (*e.g.* [29,19,17]. Decreasing cost in processing resulted in shifts in resource tradeoffs that made it feasible to consider datagram processing at line rate by the mid 1990s. A full ATM layer 3 infrastructure became unnecessary, and deployments of IP over SONET (POS – packets over SONET) began, with research into IP directly over WDM (POW – packets over wavelengths). At the same time, the important characteristics of fast packet switching technologies began to be incorporated into the Internet, for example IP switches based on the fast switch fabrics, and protocol optimisations such as MPLS.

At a high level, the architecture of a fast connectionless datagram switch depicted in Figure 6 has the same functional blocks as the fast connection oriented packet switch that was shown in Figure 2.

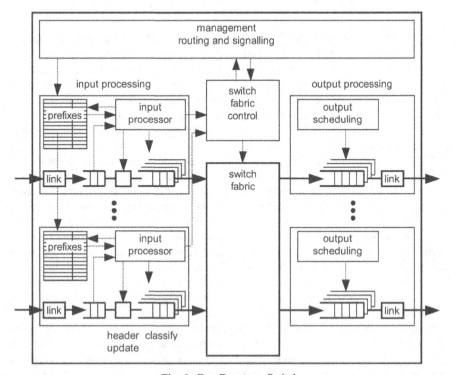

Fig. 6. Fast Datagram Switch

There is set of control software, including routing, signalling, and management, and a switch fabric core, as in a typical fast packet switch. Input and output processing are also present, and this is where the primary difference lies:

Input processing is considerably more complex, with an input processor (either a small fast RISC embedded controller or a specialised network processor) performing address lookup using a prefix table, as well as packet classification.

Output processing is also more complex, with a significant packet scheduling (including traffic shaping) to meet QOS requirement for flows, and to insure fairness among best-effort flows.

The input and output processing can either consist of custom hardware engines, or be implemented in emerging network processors. The use of network processors for this functionality opens the door to active and programmable networks, in spite or resistance by switch vendors [27].

Optical Networks. While there had long been research on optical networking, the late third generation saw significant advances in the development of optical switching components, including MEMS switch elements and the resulting optical switch fabrics; 1024×1024 research prototypes have been constructed (*e.g.* [1]). Optical switching technology provides a fast datapath, but optical logic and control circuits are beyond the ability of early fourth generation (2000s) networking. This means that all-optical *packet switching* is impractical in the near future, since the packet header cannot be decoded and processed in the optical domain. Furthermore, the switching rate of optical switch elements is relatively slow, on the order of a microsecond. Therefore, data flows must be assigned to lightpaths that are switched only infrequently. Optical burst switching [35,20] aggregates packets into bursts that can utilise the network more efficiently than circuits.

Applications. While the early third generation saw a steady increase in traffic on the Internet, primarily from educational institutions, it also saw the birth of the Web. By the mid 1990s, the exponential increase in traffic was driven by the Web, which had become ubiquitously available in universities, particularly to undergraduate students.

Applications can be classified in several ways related to their performance demands: by bandwidth aggregation, bandwidth scalability, latency requirements, and communication characteristics.

Aggregate bandwidth. An important measure of the impact of an application on the network infrastructure is the demand it places in aggregate. This is measured by the product of the *per* instance bandwidth × the number of simultaneous instantiations of the application [12]. Thus, an aggregate gigabit application might consist either of 100 simultaneous instances of a 10 Mb/s application, or 10 simultaneous instances of a 100 Mb/s application. The aggregate bandwidth of the PSTN (public switched telephone network) is generally estimated at $O(1$ Tb/s$)$ as was the bandwidth of data networks in the mid 1990s (particularly the Internet, SNA, and X.25 packet networks). While it is expected that PSTN bandwidth will remain relatively flat, the aggregate bandwidth of the Internet continues to grow dramatically with no end in sight. In the early 2000s, bundles of fibers are being laid and switches deployed that exceed 1 Tb/s.

Individual bandwidth. A single instance of an application that requires a significant fraction of the bandwidth available on a high-speed network link or high-performance end system interface can be considered a *high-speed application.* Supporting this sort of application requires high-bandwidth network infrastructure, high-speed transport protocols, and high-performance end systems. These applications clearly need to be the focus of high-speed networking research.

The bandwidth that an individual application requires is generally not a fixed quantity; most applications operate over a range of bandwidths. Thus, it is important to understand how application utility scales with available bandwidth [31]. Some applications remain structurally unchanged, becoming only faster or perceptually better; other applications have difficulty keeping pace as bandwidths scale. The bandwidth scalability of applications can be described using the following taxonomy [16]:

Bandwidth Enhanced. The application operates at various bandwidths. Although the application is functional at low bandwidths, it increases in utility given high-speed networking, and does not require fundamental restructuring. Streaming multimedia is the canonical example, because high bandwidth increases the achievable resolution and frame rate, with an increased perceptual quality to users.

Bandwidth Challenged. The application is useful at various bandwidths, but either requires substantial revision, or operates in a different way at high-speed. An example of a bandwidth challenged application is distributed computing. Some computations, such as *monte carlo* simulations, work with infrequent state exchange. As bandwidth increases, more sophisticated distributions of computation are possible, requiring greater control interaction and data exchange.

Bandwidth Enabled. The application is usable *only when* a high-bandwidth path is available. This may be dictated by particular bandwidth requirements of the application, for example the high data rates of uncompressed video for networked studio production of movies. It may also be the case that without a base bandwidth certain applications just don't make sense. Distributed scientific visualisation and collaborative CAD/CAE (computer aided design / engineering) fall into this type.

While all of these applications drive aggregate network bandwidth, is the bandwidth challenged and bandwidth enabled applications that present the most serious high-speed networking demands end-to-end and application-to-application.

Latency Characteristics. Latency is the other important characteristic of application demand, and can be characterised as best-effort, interactive, real-time, and deadline. Application utility curves, depicted in Figure 7, indicate how tolerant applications are to latency, and thus indicate the latency bound and its tolerance that must be provided by the network and end systems.

Clearly there is a range of tolerance ranging from best-effort (tolerant), through interactive (moderate) to deadline and real-time (intolerant).

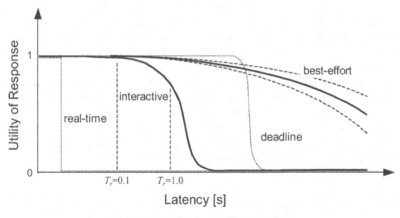

Fig. 7. Application Utility Functions

Characteristics. Additionally, applications can be categorised by characteristics class [16]: information access, telepresence, and distributed computing. Information access applications, such as the Web, are client/server with highly asymmetric bandwidth requirements and a fairly large granularity of transfers. Telepresence applications involve the exchange of information that allows users to maintain a distributed virtual presence; frequently in the form of multimedia. Telepresnece thus tends to be more symmetric in its bandwidth requirements, and individual transfers are either at small granularity or a continuous stream. Finally, distributed computing involves the distribution of computations beyond a room, and involves an arbitrary exchange of data. While requirements are highly variable on the particular computation (which is in turn designed on network capabilities), in the general case the bandwidth, latency, and synchronisation requirements can be very challenging. Other more complex application scenarios are compositions of the three core classes. For example, distance learning is a composition of telepresence (virtual classroom participation) and information access (student access to course materials).

The Web became the killer app in the mid third generation. Web browsing is an interactive information access application. It is a challenging and ubiquitous high-speed networking application, not only in aggregate, but also for each individual user browsing. Web browsing has traditionally been considered a best-effort application, but this is point, click, and wait mode. For Web browsing to meet the requirements for interactive applications (point-click-point-click), a response time in the approximate range of 100 ms to 1 second is required [25]. This latency bound drives the bandwidth requirement, especially for large web objects, such as those including embedded images. Medical and photographic-resolution images are particularly demanding.

Figure 8 [32] shows the bandwidth requirements for different types of web pages. The horizontal bands represent different types of web pages. The vertical bars indicate now much data can be transmitted over various link technologies in the 100 ms interactive response time budget, assuming the given link is the bandwidth bottleneck. Note how even modest web pages can stress analog modem and ISDN rates. As web page sizes increase with higher resolution and 3-dimensional images (allowing local rotation), bandwidth requirements increase into the Gb/s realm.

This bandwidth demand is fueled in the consumer arena by high-resolution digital cameras and printers, coupled with the desire to deliver of digital photographs on the Web. Note that this analysis only considers the propagation delay, which assumes the entire 100 ms of response time budget can be used for data transmission. Client, server, and network node delays also contribute to the end-to-end interapplication delay, and may consume a significant portion of the latency budget, requiring even higher link bandwidths. Servers outside a 100 ms propagation radius (around 5000 km) cannot be accessed within this latency bound at all with direct request-response techniques; this motivates techniques that masking latency.

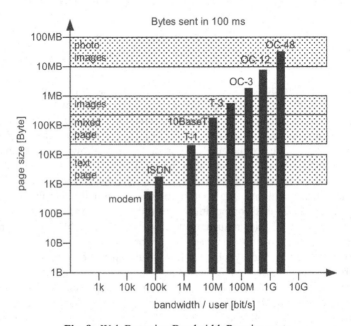

Fig. 8. Web Browsing Bandwidth Requirements

Application matching the network. Significant performance gains are possible if applications are aware of the underlying network data structures and control mechanisms; matching them can dramatically reduce the control overhead and data transformation. The PROTOCOL LAYERING PRINCIPLE introduced previously indicates that layering as an abstraction need not lead to poor implementation. ILP is one embodiment of this principle. Another is *application layer framing* (ALF), which is a technique that allows the application to more directly adapt network protocol formats and data units [6]. This reduces the overhead in protocol encapsulation, decapsulation, fragmentation, and reassembly. There are benefits in matching in the control plane as well. Unfortunately, HTTP is an application layer transaction protocol hacked onto a connection-oriented transport protocol designed for long-lived data transfers. While there were attempts to modify TCP for transactions [2] they were not deployed, due to security flaws in TCP connection management. Furthermore, there is no attempt to match the structure of Web pages to the underlying protocol data units. Thus the Web benefits neither from a native transaction protocol, nor from ALF.

Distributed Data. The way in which data is structured and distributed across the network can have a profound influence on interapplication delay, and thus on application-to-application performance.

DISTRIBUTED DATA PRINCIPLE: *Distributed applications should select and organise the data they exchange to minimise the amount and latency of data transferred and to allow incremental processing of data.*

Unfortunately, there has been little tangible evidence of this principle in widely deployed applications. Many applications perform poorly and *seem* to need high-speed networks simply because they are poorly designed and partitioned. The Web is again an example of this problem; web content is generally not structured with performance in mind. Rather than organising data into easily transferable, displayable, and cacheable units, web page designers use authoring tools that have no cognisance of this; the overuse and misuse of dynamic content is a reflection of this problem. Furthermore, administrative and policy decisions are frequently are at odds with performance (recall the LIMITING CONSTRAINTS axiom from Section 2). While some of these limiting constraints have practical justification, they are frequently not balanced against the needs of high-speed applications. The way in which banner advertisements are implemented is an example of this problem.

Application Adaptation and Latency Masking. In an attempt to adapt to constraints on latency (primarily due to the speed-of-light over long distances) and limited bandwidth, applications can adapt to mask these effects. This depends on network feedback, and may benefit from user control [28]. Mirroring and caching are the canonical techniques to mask latency and reduce aggregate bandwidth, and this became an intense area of research in the late 1990s. There are limits to the benefits of these techniques, however, and the next step are anticipatory techniques that prefetch and preload, in an attempt to reduce response time for pages that are not yet cached. Simple examples are to prefetch pages hyperlinked in the page just requested [30] and to push preload based on user profile information located on a server. Intelligent rate adaptation and layered coding help applications to gracefully degrade as bandwidth becomes constrained.

3.4 Fourth Generation

The first decade of the new millennium inaugurates a new network generation, which will be characterised by orders of magnitude increase in network scale, and by the ubiquity of mobile wireless computing devices. The third generation was largely a wired network; the fourth generation will be largely wireless at the edges, with access to a high-speed optical backbone infrastructure, including optical switches. In the extreme, one can envisage a network that consists *only* of wireless access to an optical backbone, although in practice copper wire will be in use for a very long time.

Advances in the interface between biological and microelectronic interfaces, and the benefits of applying micro- and macro-biotic behaviour and organisation techniques to networks may begin to emerge, or may be a characteristic of a later fifth generation.

Ubiquitous computing, smart spaces, and sensor network research suggest that individual users will carry tens to hundreds of networked devices. These will perhaps include a personal node for computing and communications to other individuals, and a wireless network of wearable I/O devices and sensors (environmental and biological). Rooms and vehicles will consist of perhaps thousands of embedded sensors and networked computing platforms. Thus we need to consider teranode and petanode networks. There are profound implications to high-speed networking. The aggregate wireless bandwidth demanded will vastly exceed the capacity of the shared medium using third generation protocols and techniques, and the highly dynamic nature will stress routing and control protocols. The ability to manage power among the massive number of autonomous wireless devices, and to do high-speed networking where power is a constraint will be a major challenge.

We will see not only end systems, the sole purpose of which is networking, but also a blurring of functionality between end systems and network nodes. In mobile networking, many devices will serve both as application platforms and as switches or routers.

While the capacity in processing, memory, and bandwidth will dramatically increase, resource tradeoffs will continue to shift. If the shifts are significant (for example several orders of magnitude increase in only one of processing, memory, or bandwidth), the future of high-speed networking will be drastically different.

The relative decrease in the cost of processing enabled the field of active networking, which may play a significant role in the fourth generation. We note that speed-of-light latency will continue to be a significant challenge, and increasingly so as we begin to build the Interplanetary Internet, initially for the Mars missions, but with an eye toward the Jupiter system.

4   The Future of High-Speed Networking as a Discipline

High-speed networking has become a mature discipline, to the point that everything has *some* aspect of high-speed networking, and nothing is *only* high-speed networking. In the late 1990s, this seemed like a reasonable state of affairs.

Unfortunately, the decline in high-speed networking as a distinct discipline seems to have lead to the situation where nobody is looking after the performance of the *entire* network as a *system of systems*. At best, component manufacturers are building high-performance subsystems (such as fast IP switches), but typically service providers deploy them in a haphazard manner to barely stay ahead of the demand curve. At worst, network providers are working at odds with one another deploying bad topologies with complex and irrational peering points that obscure performance. ASPs are deploying hacks and middleboxes without regard to the overall performance of the Internet.

Active and programmable networks may provide the mechanisms to evolve the network in a systematic manner; switches that contain network processors may allow this to happen in a rational manner in spite of switch vendors that do not wish to open their boxes, and network service providers that can't see beyond the next bandwidth capacity planning cycle.

At best, high-speed networking is in a rut [13]; at worst it has been fragmented and absorbed into the mainstream. As long as there is a community of people deeply interested in high-speed networking, there is hope. Whether this translates into an effort to restore order and performance to a chaotic network remains to be seen.

References

1. David J. Bishop, C. Randy Giles, and Gary P. Austin, "The Lucent LambdaRouter: MEMS Technology of the Future Here Today", *IEEE Communications*, vol.40 #3, IEEE, New York NY US, Mar. 2002, pp. 75–79
2. Robert Braden, *Extending TCP for Transactions – Concepts*, RFC 1379, Nov. 1992.
3. Vinton G. Cerf and Edward Cain, "The DoD Internet Architecture Model", *Computer Networks*, vol.7 #5, Elsevier Science / North-Holland, Amsterdam NL, Oct. 1983, pp. 307–318.
4. Greg Chesson, "XTP/PE Design Considerations", in *Protocols for High-Speed Networks*, IFIP PfHSN'89 (Zürich CH), May 1989, Harry Rudin and Robin Williamson editors, Elsevier / North-Holland, Amsterdam NL, 1989, pp. 27–33.
5. David D. Clark, Van Jacobson, John Romkey, and Howard Salwen, "An Analysis of TCP Processing Overhead", *IEEE Communications*, vol.27 #.6, IEEE, New York NY US, June 1989, pp. 23–29.
6. David D. Clark and David L. Tennenhouse, "Architectural Considerations for a New Generation of Protocols", *Proceedings ACM SIGCOMM'90* (Philadelpha PA US), *Computer Communication Review*, vol.20 #4, ACM, New York NY US, Sep. 1990, pp. 200–208.
7. Bruce S. Davie, "A Host–Network Interface Architecture for ATM", *Proceedings of ACM SIGCOMM'91* (Zürich CH), *Computer Communication Review*, vol.21 #4, ACM, New York NY US, Sep. 1991, pp. 307–315.
8. Gary S. Delp, Adarshpal S. Sethi, and David J. Farber, "An Analysis of Memnet: An Experiment in High-Speed Shared-Memory Local Networking", *Proceedings of ACM SIGCOMM'88* (Stanford CA US), *Computer Communication Review*, vol.18 #4, ACM, New York NY US, Aug. 1988, pp. 165–174.
9. David C. Feldmeier, "A Framework of Architectural Concepts for High-Speed Communications Systems", *IEEE Journal on Selected Areas in Communications*, vol.11 #4, IEEE, New York NY US, May 1993, pp. 480–488.
10. Zygmunt Haas, "A Communication Architecture for High Speed Networking", *Proceedings of IEEE INFOCOM'90* (San Francisco CA US), IEEE, New York NY US, June 1990, pp. 433–441.
11. Van Jacobson, Robert Braden, and David A. Borman, *TCP Extensions for High Performance*, RFC 1323 (standards track), May 1992.
12. J. Bryan Lyles, Ira Richer, and James P.G. Sterbenz, "Applications Enabling the Wide Scale Deployment of Gigabit Networks" (editorial), *IEEE Journal on Selected Areas in Communications*, vol.13 #5, IEEE, New York NY US, June 1995, pp.765–767.
13. J. Bryan Lyles, keynote address, Protocols for High-Speed Networks, Berlin DE, Apr. 2002.

14. John M. McQuillan and David Walden, "The ARPA Network Design Decisions", *Computer Networks*, vol.1 #5, North-Holland, Amsterdam NL, Aug. 1977, pp. 243–289.

15. Robert M. Metcalfe and David R. Boggs, "Ethernet: Distributed Packet Switching for Local Computer Networks", *Communications of the ACM*, vol.19 #5, ACM, New York NY, Jul. 1976, pp. 395–404.

16. Craig Partridge editor, *Report of the ARPA/NSF Workshop on Research in Gigabit Networking*, Washington DC, Jul. 1994, available from http://www.cise.nsf.gov/anir/giga/craig.txt.

17. Craig Partridge, Philip P. Carvey, Ed Burgess, Isidro Castineyra, Tom Clarke, Lise Graham, Michael Hathaway, Phil Herman, Allen King, Steve Kolhami, Tracy Ma, John Mcallen, Trevor Mendez, Walter C. Milliken, Ronald Pettyjohn, John Rokosz, Joshua Seeger, Michael Sollins, Steve Storch, Benjamin Tober, Gregory D. Troxel, David Waitzman, and Scott Winterble, "A 50-Gb/s IP Router", IEEE/ACM Transactions on Networking, vol.6 #3, IEEE / ACM, New York NY US, Jun. 1998, pp. 237–248.

18. Gurudatta M. Parulkar and Jonathan S. Turner, "Towards a Framework for High-Speed Communication in a Heterogeneous Networking Environment", *IEEE Network*, vol.4 #2, IEEE, New York NY US, Mar. 1990, pp. 19–27.

19. Guru Parulkar, Douglas C. Schmidt, and Jonathan S. Turner, "a^{1t}Pm: A Strategy for Integrating IP with ATM", *Proceedings of ACM SIGCOMM'95*, (Cambridge MA US), *Computer Communication Review*, vol.25 #4, ACM, New York NY US, Aug. 1995, pp. 49–57.

20. Chunming Qiao and Myungsik Yoo, "Optical Burst Switching – A New Paradigm for an Optical Internet", *Journal of High Speed Networks*, vol.8 #1, 1999, pp. 69–84.

21. John S. Quarterman, *The Matrix: Computer Networks and Conferencing Systems Worldwide*, Digital Press, Maynard MA US, 1989.

22. J.H. Saltzer, D.P. Reed, and D.D. Clark, "End-to-end Arguments in System Design," *Proceedings of the Second International Conference on Distributed Computing Systems (ICDCS)*, IEEE, New York NY US, 1981, pp. 509–512, also *ACM Transactions on Computer Systems*, vol.2 #4, ACM, New York NY US, Nov. 1984, 227–288.

23. James P.G. Sterbenz and Gurudatta M. Parulkar, "Axon: A Distributed Communication Architecture for High-Speed Networking", *Proceedings of IEEE INFOCOM'90* (San Francisco CA US), June 1990, pp 415–425.

24. James P.G. Sterbenz and Gurudatta M. Parulkar, "Axon Network Virtual Storage for High Performance Distributed Applications", *Proceedings of 10th International Conference on Distributed Computing Systems ICDCS* (Paris FR), IEEE, New York NY US, June 1990, pp 484–492.

25. James P.G. Sterbenz, "Protocols for High Speed Networks: Life After ATM?", *Protocols for High Speed Networks IV*, IFIP/IEEE PfHSN'94 (Vancouver BC CA), Aug. 1994, Gerald Neufeld and Mabo Ito, editors, Chapman & Hall, London UK / Kluwer Academic Publishers, Norwell MA US, 1995, pp. 3–18.

26. James P.G. Sterbenz and Joseph D. Touch, High-Speed Networking: A Systematic Approach to High-Bandwidth Low-Latency Communication, John Wiley, New York NY US, 2001.

27. James P.G. Sterbenz, "Intelligence in Future Broadband Networks: Challenges and Opportunities in High-Speed Active Networking", *Proceedings of IEEE International Zürich Seminar on Broadband Communications IZS 2002* (Zürich CH), IEEE, New York, Feb. 2002, pp. 2-1-2-7.

28. James P.G. Sterbenz, Rajesh Krishnan, and Tushar Saxena, *Latency Aware Information Acces with User Directed Handling of Cache Misses: Web VADE MECUM*, http://www.ir.bbn.com/projects/wvm.

29. Ahmed Tantawy and Martina Zitterbart, "Multiprocessing in High Performance IP Routers", *Protocols for High Speed Networks III*, IFIP PfHSN'92 (Stockholm SE), May

1992, Per Gunningberg, Björn Perhson, and Stephen Pink editors, Elsevier / North-Holland, Amsterdam NL, 1993, pp. 235–254.

30. Joseph D. Touch, "Parallel Communication" *Proceedings of INFOCOM'93* (San Francisco CA US), IEEE, New York NY US, Mar. 1993, pp. 506–512.

31. Joseph D. Touch, "Defining 'High Speed' Protocols : Five Challenges and an Example That Survives the Challenges", *IEEE Journal on Selected Areas in Communications*, vol.13, #5, IEEE, New York NY US, June 1995, pp. 828–835.

32. Joseph D. Touch., "High Performance Web", animation session, *Protocols for High-Speed Network*, IFIP/IEEE, PfHSN'96 (Sophia-Antipolis, FR), Oct. 1996.

33. C. Brandan S. Traw and Jonathan M. Smith, "Hardware/Software Organization of a High-Performance ATM Host Interface", *IEEE Journal on Selected Areas in Communications*, vol.11 #2, IEEE, New York NY US, Feb. 1993, pp.228–239.

34. Jonathan S. Turner, "Design of an Integrated Services *Packet* Network", *IEEE Journal on Selected Areas in Communications*, vol.SAC-4 #8, IEEE, New York NY US, Nov. 1986, pp. 1373–1380.

35. Jonathan S. Turner, "Terabit Burst Switching", *Journal of High Speed Networks*, vol.8 #1, IOS Press, Amsterdam NL, 1999, pp. 3–16.

36. Richard W. Watson and Sandy A. Mamrak, "Gaining Efficiency in Transport Services by Appropriate Design and Implementation Choices", *ACM Transactions on Computer Systems*, vol.5 #2, May 1987, pp. 97–120.

37. C.M. Woodside, K. Ravinadran, and R.G. Franks, "The Protocol Bypass Concept for High Speed OSI Data Transfer." *Protocols for High-Speed Networks II*, IFIP PfHSN'1990 (Palo Alto CA US), Oct. 1990, Marjory Johnson editor, Elsevier / North-Holland, Amsterdam NL, 1991, pp. 107–122.

Author Index

Abrahamsson, H.   117
Ajmone Marsan, M.   100
Alessio, E.   100

Blondia, C.   211

Casals, O.   211
Chan, C.-T.   180
Chen, Y.-C.   180
Cleyn, P. De   211
Cnodder, S. De   164
Cohen, R.   84
Császár, A.   17

Dabran, I.   84

Feng, W.-c.   69
Fisk, M.   69

Gardner, M.   69
Garetto, M.   100
Grieco, L.A.   130
Gunningberg, P.   50

Hagsand, O.   117

Karagiannis, G.   17
Karlsson, G.   147

Liebeherr, J.   192
Lo Cigno, R.   100

Marquetant, Á.   17
Marsh, I.   117

Martin, R.   35
Mascolo, S.   130
Menth, M.   35
Meo, M.   100
Mysore, J.P.   1

Needham, M.   1

Ossipov, E.   147

Partain, D.   17
Patek, S.   192
Pelsser, C.   164
Pop, O.   17

Rexhepi, V.   17

Schrodi, K.J.   229
Sterbenz, J.P.G.   227, 243
Szabó, R.   17

Takács, A.   17
Touch, J.D.   228
Tseng, W.-C.   180

Venkitaraman, N.   1
Voigt, T.   50

Wang, H.   192
Wang, J.   192
Wang, P.-C.   180
Weigle, E.   69
Westberg, L.   17
Willems, G.   211

Lecture Notes in Computer Science

For information about Vols. 1–2248
please contact your bookseller or Springer-Verlag

Vol. 2248: C. Boyd (Ed.), Advances in Cryptology – ASIACRYPT 2001. Proceedings, 2001. XI, 603 pages. 2001.

Vol. 2249: K. Nagi, Transactional Agents. XVI, 205 pages. 2001.

Vol. 2250: R. Nieuwenhuis, A. Voronkov (Eds.), Logic for Programming, Artificial Intelligence, and Reasoning. Proceedings, 2001. XV, 738 pages. 2001. (Subseries LNAI).

Vol. 2251: Y.Y. Tang, V. Wickerhauser, P.C. Yuen, C.Li (Eds.), Wavelet Analysis and Its Applications. Proceedings, 2001. XIII, 450 pages. 2001.

Vol. 2252: J. Liu, P.C. Yuen, C. Li, J. Ng, T. Ishida (Eds.), Active Media Technology. Proceedings, 2001. XII, 402 pages. 2001.

Vol. 2253: T. Terano, T. Nishida, A. Namatame, S. Tsumoto, Y. Ohsawa, T. Washio (Eds.), New Frontiers in Artificial Intelligence. Proceedings, 2001. XXVII, 553 pages. 2001. (Subseries LNAI).

Vol. 2254: M.R. Little, L. Nigay (Eds.), Engineering for Human-Computer Interaction. Proceedings, 2001. XI, 359 pages. 2001.

Vol. 2255: J. Dean, A. Gravel (Eds.), COTS-Based Software Systems. Proceedings, 2002. XIV, 257 pages. 2002.

Vol. 2256: M. Stumptner, D. Corbett, M. Brooks (Eds.), AI 2001: Advances in Artificial Intelligence. Proceedings, 2001. XII, 666 pages. 2001. (Subseries LNAI).

Vol. 2257: S. Krishnamurthi, C.R. Ramakrishnan (Eds.), Practical Aspects of Declarative Languages. Proceedings, 2002. VIII, 351 pages. 2002.

Vol. 2258: P. Brazdil, A. Jorge (Eds.), Progress in Artificial Intelligence. Proceedings, 2001. XII, 418 pages. 2001. (Subseries LNAI).

Vol. 2259: S. Vaudenay, A.M. Youssef (Eds.), Selected Areas in Cryptography. Proceedings, 2001. XI, 359 pages. 2001.

Vol. 2260: B. Honary (Ed.), Cryptography and Coding. Proceedings, 2001. IX, 416 pages. 2001.

Vol. 2261: F. Naumann, Quality-Driven Query Answering for Integrated Information Systems. X, 166 pages. 2002.

Vol. 2262: P. Müller, Modular Specification and Verification of Object-Oriented Programs. XIV, 292 pages. 2002.

Vol. 2263: T. Clark, J. Warmer (Eds.), Object Modeling with the OCL. VIII, 281 pages. 2002.

Vol. 2264: K. Steinhöfel (Ed.), Stochastic Algorithms: Foundations and Applications. Proceedings, 2001. VIII, 203 pages. 2001.

Vol. 2265: P. Mutzel, M. Jünger, S. Leipert (Eds.), Graph Drawing. Proceedings, 2001. XV, 524 pages. 2002.

Vol. 2266: S. Reich, M.T. Tzagarakis, P.M.E. De Bra (Eds.), Hypermedia: Openness, Structural Awareness, and Adaptivity. Proceedings, 2001. X, 335 pages. 2002.

Vol. 2267: M. Cerioli, G. Reggio (Eds.), Recent Trends in Algebraic Development Techniques. Proceedings, 2001. X, 345 pages. 2001.

Vol. 2268: E.F. Deprettere, J. Teich, S. Vassiliadis (Eds.), Embedded Processor Design Challenges. VIII, 327 pages. 2002.

Vol. 2269: S. Diehl (Ed.), Software Visualization. Proceedings, 2001. VIII, 405 pages. 2002.

Vol. 2270: M. Pflanz, On-line Error Detection and Fast Recover Techniques for Dependable Embedded Processors. XII, 126 pages. 2002.

Vol. 2271: B. Preneel (Ed.), Topics in Cryptology – CT-RSA 2002. Proceedings, 2002. X, 311 pages. 2002.

Vol. 2272: D. Bert, J.P. Bowen, M.C. Henson, K. Robinson (Eds.), ZB 2002: Formal Specification and Development in Z and B. Proceedings, 2002. XII, 535 pages. 2002.

Vol. 2273: A.R. Coden, E.W. Brown, S. Srinivasan (Eds.), Information Retrieval Techniques for Speech Applications. XI, 109 pages. 2002.

Vol. 2274: D. Naccache, P. Paillier (Eds.), Public Key Cryptography. Proceedings, 2002. XI, 385 pages. 2002.

Vol. 2275: N.R. Pal, M. Sugeno (Eds.), Advances in Soft Computing – AFSS 2002. Proceedings, 2002. XVI, 536 pages. 2002. (Subseries LNAI).

Vol. 2276: A. Gelbukh (Ed.), Computational Linguistics and Intelligent Text Processing. Proceedings, 2002. XIII, 444 pages. 2002.

Vol. 2277: P. Callaghan, Z. Luo, J. McKinna, R. Pollack (Eds.), Types for Proofs and Programs. Proceedings, 2000. VIII, 243 pages. 2002.

Vol. 2278: J.A. Foster, E. Lutton, J. Miller, C. Ryan, A.G.B. Tettamanzi (Eds.), Genetic Programming. Proceedings, 2002. XI, 337 pages. 2002.

Vol. 2279: S. Cagnoni, J. Gottlieb, E. Hart, M. Middendorf, G.R. Raidl (Eds.), Applications of Evolutionary Computing. Proceedings, 2002. XIII, 344 pages. 2002.

Vol. 2280: J.P. Katoen, P. Stevens (Eds.), Tools and Algorithms for the Construction and Analysis of Systems. Proceedings, 2002. XIII, 482 pages. 2002.

Vol. 2281: S. Arikawa, A. Shinohara (Eds.), Progress in Discovery Science. XIV, 684 pages. 2002. (Subseries LNAI).

Vol. 2282: D. Ursino, Extraction and Exploitation of Intensional Knowledge from Heterogeneous Information Sources. XXVI, 289 pages. 2002.

Vol. 2283: T. Nipkow, L.C. Paulson, M. Wenzel, Isabelle/HOL. XIII, 218 pages. 2002.

Vol. 2284: T. Eiter, K.-D. Schewe (Eds.), Foundations of Information and Knowledge Systems. Proceedings, 2002. X, 289 pages. 2002.

Vol. 2285: H. Alt, A. Ferreira (Eds.), STACS 2002. Proceedings, 2002. XIV, 660 pages. 2002.

Vol. 2286: S. Rajsbaum (Ed.), LATIN 2002: Theoretical Informatics. Proceedings, 2002. XIII, 630 pages. 2002.

Vol. 2287: C.S. Jensen, K.G. Jeffery, J. Pokorny, Saltenis, E. Bertino, K. Böhm, M. Jarke (Eds.), Advances in Database Technology – EDBT 2002. Proceedings, 2002. XVI, 776 pages. 2002.

Vol. 2288: K. Kim (Ed.), Information Security and Cryptology – ICISC 2001. Proceedings, 2001. XIII, 457 pages. 2002.

Vol. 2289: C.J. Tomlin, M.R. Greenstreet (Eds.), Hybrid Systems: Computation and Control. Proceedings, 2002. XIII, 480 pages. 2002.

Vol. 2291: F. Crestani, M. Girolami, C.J. van Rijsbergen (Eds.), Advances in Information Retrieval. Proceedings, 2002. XIII, 363 pages. 2002.

Vol. 2292: G.B. Khosrovshahi, A. Shokoufandeh, A. Shokrollahi (Eds.), Theoretical Aspects of Computer Science. IX, 221 pages. 2002.

Vol. 2293: J. Renz, Qualitative Spatial Reasoning with Topological Information. XVI, 207 pages. 2002. (Subseries LNAI).

Vol. 2294: A. Cortesi (Ed.), Verification, Model Checking, and Abstract Interpretation. Proceedings, 2002. VIII, 331 pages. 2002.

Vol. 2295: W. Kuich, G. Rozenberg, A. Salomaa (Eds.), Developments in Language Theory. Proceedings, 2001. IX, 389 pages. 2002.

Vol. 2296: B. Dunin-Kęplicz, E. Nawarecki (Eds.), From Theory to Practice in Multi-Agent Systems. Proceedings, 2001. IX, 341 pages. 2002. (Subseries LNAI).

Vol. 2297: R. Backhouse, R. Crole, J. Gibbons (Eds.), Algebraic and Coalgebraic Methods in the Mathematics of Program Construction. Proceedings, 2000. XIV, 387 pages. 2002.

Vol. 2299: H. Schmeck, T. Ungerer, L. Wolf (Eds.), Trends in Network and Pervasive Computing – ARCS 2002. Proceedings, 2002. XIV, 287 pages. 2002.

Vol. 2300: W. Brauer, H. Ehrig, J. Karhumäki, A. Salomaa (Eds.), Formal and Natural Computing. XXXVI, 431 pages. 2002.

Vol. 2301: A. Braquelaire, J.-O. Lachaud, A. Vialard (Eds.), Discrete Geometry for Computer Imagery. Proceedings, 2002. XI, 439 pages. 2002.

Vol. 2302: C. Schulte, Programming Constraint Services. XII, 176 pages. 2002. (Subseries LNAI).

Vol. 2303: M. Nielsen, U. Engberg (Eds.), Foundations of Software Science and Computation Structures. Proceedings, 2002. XIII, 435 pages. 2002.

Vol. 2304: R.N. Horspool (Ed.), Compiler Construction. Proceedings, 2002. XI, 343 pages. 2002.

Vol. 2305: D. Le Métayer (Ed.), Programming Languages and Systems. Proceedings, 2002. XII, 331 pages. 2002.

Vol. 2306: R.-D. Kutsche, H. Weber (Eds.), Fundamental Approaches to Software Engineering. Proceedings, 2002. XIII, 341 pages. 2002.

Vol. 2307: C. Zhang, S. Zhang, Association Rule Mining. XII, 238 pages. 2002. (Subseries LNAI).

Vol. 2308: I.P. Vlahavas, C.D. Spyropoulos (Eds.), Methods and Applications of Artificial Intelligence. Proceedings, 2002. XIV, 514 pages. 2002. (Subseries LNAI).

Vol. 2309: A. Armando (Ed.), Frontiers of Combining Systems. Proceedings, 2002. VIII, 255 pages. 2002. (Subseries LNAI).

Vol. 2310: P. Collet, C. Fonlupt, J.-K. Hao, E. Lutton, M. Schoenauer (Eds.), Artificial Evolution. Proceedings, 2001. XI, 375 pages. 2002.

Vol. 2311: D. Bustard, W. Liu, R. Sterritt (Eds.), Soft-Ware 2002: Computing in an Imperfect World. Proceedings, 2002. XI, 359 pages. 2002.

Vol. 2312: T. Arts, M. Mohnen (Eds.), Implementation of Functional Languages. Proceedings, 2001. VII, 187 pages. 2002.

Vol. 2313: C.A. Coello Coello, A. de Albornoz, L.E. Sucar, O.Cairó Battistutti (Eds.), MICAI 2002: Advances in Artificial Intelligence. Proceedings, 2002. XIII, 548 pages. 2002. (Subseries LNAI).

Vol. 2314: S.-K. Chang, Z. Chen, S.-Y. Lee (Eds.), Recent Advances in Visual Information Systems. Proceedings, 2002. XI, 323 pages. 2002.

Vol. 2315: F. Arhab, C. Talcott (Eds.), Coordination Models and Languages. Proceedings, 2002. XI, 406 pages. 2002.

Vol. 2316: J. Domingo-Ferrer (Ed.), Inference Control in Statistical Databases. VIII, 231 pages. 2002.

Vol. 2317: M. Hegarty, B. Meyer, N. Hari Narayanan (Eds.), Diagrammatic Representation and Inference. Proceedings, 2002. XIV, 362 pages. 2002. (Subseries LNAI).

Vol. 2318: D. Bošnački, S. Leue (Eds.), Model Checking Software. Proceedings, 2002. X, 259 pages. 2002.

Vol. 2319: C. Gacek (Ed.), Software Reuse: Methods, Techniques, and Tools. Proceedings, 2002. XI, 353 pages. 2002.

Vol. 2322: V. Mařík, O. Štěpánková, H. Krautwurmová, M. Luck (Eds.), Multi-Agent Systems and Applications II. Proceedings, 2001. XII, 377 pages. 2002. (Subseries LNAI).

Vol. 2324: T. Field, P.G. Harrison, J. Bradley, U. Harder (Eds.), Computer Performance Evaluation. Proceedings, 2002. XI, 349 pages. 2002.

Vol. 2329: P.M.▲. Sloot, C.J.K. Tan, J.J. Dongarra, A.G. Hoekstra (Eds.), Computational Science – ICCS 2002. Proceedings, Part I. XLI, 1095 pages. 2002.

Vol. 2330: P.M.A. Sloot, C.J.K. Tan, J.J. Dongarra, A.G. Hoekstra (Eds.), Computational Science – ICCS 2002. Proceedings, Part II. XLI, 1115 pages. 2002.

Vol. 2331: P.M.A. Sloot, C.J.K. Tan, J.J. Dongarra, A.G. Hoekstra (Eds.), Computational Science – ICCS 2002. Proceedings, Part III. XLI, 1227 pages. 2002.

Vol. 2332: L. Knudsen (Ed.), Advances in Cryptology – EUROCRYPT 2002. Proceedings, 2002. XII, 547 pages. 2002.

Vol. 2334: G. Carle, M. Zitterbart (Eds.), Protocols for High Speed Networks. Proceedings, 2002. X, 267 pages. 2002.